BRIDGEWATER
The Canal Duke, 1736–1803

Francis Egerton, Baron Ellesmere, Marquis of Brackley, eighth Earl, and third and last Duke of Bridgewater, b. 21 May 1736, d. 8 March 1803: from the miniature by Richard Crosse (*by courtesy of the Tatton estate*)

BRIDGEWATER
The Canal Duke, 1736–1803

HUGH MALET

All these things considered, it is not too much to say that the Duke of Bridgewater was the greatest benefactor the great towns of Liverpool and Manchester ever saw.

Edward Kirk, *A Guide to Worsley, Historical & Topographical*, 1870

He did, perhaps, more to promote the prosperity of this country than all the Dukes, Marquises and Earls put together.

Sir Spencer Walpole

Hendon Publishing : Nelson

First edition 1961

Revised edition 1977
by
Manchester University Press
Second impression 1990
by
Hendon Publishing Co. Ltd.

ISBN 0 86067 136 4

Printed in Great Britain by
Athenaeum Press Ltd.
Newcastle upon Tyne NE4 6T

in association with
Richard Netherwood Limited
for
Hendon Publishing Co. Ltd.
Hendon Mill, Nelson, Lancs.

Contents

List of illustrations

Acknowledgements

At the beginning of the dedication of the *In Parenthesis* the poet stated succinctly, 'This writing is for my friends in mind.' Similarly I acknowledge with gratitude my debt to those who have so long guided this work among the shoals and perils of research. Dr W. H. Chaloner, Professor of Economic History at Manchester University, and the pioneer in this field, has been endlessly patient with my many enquiries. Professor H. P. White, of Salford University, has leavened the work with his extensive knowledge of the geography and economics of transport history. Mr Charles Hadfield, the canal historian, has added his encyclopaedic knowledge and corrections to the text. Mr Frank Mullineux, the distinguished local historian, has guided me with much wisdom, as well as local lore, while Mr F. C. Mather of Southampton University has helped me with the later section. The generous decision of John Sutherland Egerton, the sixth Duke, and of the Duchess of Sutherland, to make their Bridgewater archives available has helped to confirm many of my previous findings in this field of research, and has also opened unexplored vistas of economic history.

To Mr I. H. C. Powell and Dr I. F. W. Beckett I am indebted for guidance and advice leading to a generous subvention, granted by the Academic Publishing Committee of the University of Salford. I would have found it difficult indeed to carry out so much research in many different places without the grants from the Fraser Trust and from the Directors of the Manchester Ship Canal Company. For the first I have to thank my former colleague, the Rev. J. K. Byrom, lately Warden of Brasted Theological College, while the second was obtained through the kind offices of Mr A. Hayman, Manager of the Bridgewater Department, who has also guided me with valuable information.

In the notes I have acknowledged the kind help of a number of people, but would here particularly mention Dr D. E. Owen, formerly Director of the

Manchester Museum, the Lancashire historian Mr J. J. Bagley, Mr John Garner, Director of Fine Arts at Salford University, Mr C. E. Makepeace, Mr. Alan Jeffery and Mr S. R. Broadbridge, Mr G. S. Darlow, Miss Moira Leggat, the Rev. H. E. S. Meanley, Mr and Mrs Arthur Lucas of Woolmers Park, Mr Peter Walne, Mr Randall J. Le Boeuf, Jnr., Mr M. W. Moss, Mr Frank Sharman, Mr H. M. G. Concannon, Mr Alan Smith, Mr John Shirt and Mr Anthony Frankland.To the staff of Tatton Park I am indebted for much help over illustrations.

I also gratefully acknowledge the help of Miss Kathleen Shawcross in assembling a difficult manuscript, and my wife's patient encouragement.

Abbreviations

BES	Bridgewater Estate Company Archives, John Massey's Account Books
CAL	Calendared archives
CNW	Charles Hadfield and Gordon Biddle, *Canals of North West England*, 2 vols, (Newton Abbot, 1970)
CRO	County Record Office
DNB	*Dictionary of National Biography*
EM	Edith Malley (Mrs E. Brickell), the Financial Administration of the Bridgewater Estate, 1780–1800. Unpublished MA thesis, Manchester University, 1929
HLRO	House of Lords Record Office
HPT	History of Parliament Trust, London University
JHC	Journal of the House of Commons
JHL	Journal of the House of Lords
MLH	Manchester Central Library, Local History Library
SEM	Sutherland Estate papers, relating to Francis Egerton, Mertoun, Roxburghshire
SH	Strachan Holme, *Family Chronicles*, sub-section *The Life of the Most Noble Francis Egerton, Baron Ellesmere, Marquis of Brackley, 6th Earl and 3rd (and last) Duke of Bridgewater* . . ., unpublished, n.d. (*c.* 1920), located in SEM. A series of notes and extracts divided into sections
SS	Samuel Smiles, *Lives of the Engineers*, vol. I (1961). Page references are to the 1904 edn.
TLC	Transactions of the Lancashire and Cheshire Antiquarian Society

Historical note

The Canal Duke, my first biography of Francis Egerton, was published in 1961, almost at the beginning of the Canal Revival. Although it had been designed to be read more for pleasure than for improvement, its historical content aroused widespread interest, resulting in many years of correspondence with scholars and others linked with the Bridgewater achievement. Material from some of those valuable contributions has been included in the following pages, but it is the documentary evidence which has added the main substance to the story.

This considerable corpus of new archive material deserved, it was felt, the present and fundamentally different book, which has aimed at a rather more academic approach without, one hopes, losing that sense of high endeavour and adventurous innovation which marked all the duke's activities. Although most of the original strands of argument expounded in *The Canal Duke* have been substantially vindicated by the new documentary evidence, it may be a help to the general reader if I summarise some of those historical conclusions in the following note.

Biographers can too readily eulogise their subjects, but on the other hand justice has to be done, and the derogatory myth of the duke's meanness in underpaying employees like Brindley has at last been completely exploded; on the contrary, Francis Egerton should now be rehabilitated as one of the major benefactors and most liberal masters of his age, or of any other.

The more tenacious myth purporting to advance the candidature of James Brindley as the inspired genius behind the Bridgewater canal long continued to find a happy acceptance among some historians, but we can now muster a host of witnesses whose evidence runs counter, including, among others, Sir Joseph Banks, Strachan Holme, the Rev. J. Fenton, Eliza Meteyard, the Rev. F. H. Egerton and the Northampton account books. A proper study of these sources must surely relegate Brindley's brief if sometimes useful contribution to the Bridgewater story to its proper perspective at last.

There is, perhaps, a need for writers in this field to guard against oversimplifying a canal achievement by assuming that evidence of some surveying and engineering work makes it safe to attribute a waterway entirely to the kind of 'cult figure' of the rise from rags to riches that Brindley had become; for where more precise evidence is lacking this kind of solution can prove unsatisfactory. There was infinitely more to canal work than merely surveying it, and certainly the engineering problems on the Bridgewater were formidable, but the duke appreciated that the chief challenge lay, intially, with the parliamentary opposition, so it is there, in the salons and lobbies of London, that we find him most often from 1757 to 1762. That is why George Moseley Lander, who was descended from both the Gilberts and the Bills, could write emphatically in 1883 (much as Arthur Young had done earlier) of 'the great duke of Bridgewater, the suggestor and *inventor* of the Bridgewater canal', but these and other hints were rapidly submerged beneath the popularity of Smiles's biased and inaccurate but persuasive prose, from which subsequent writers have quarried so avidly. Nowadays it is more clear that Brindley & Co.'s purposeful pursuit of publicity was in marked contrast with the duke's disdain for it.

Although the circumstances were in some ways so different, I also felt that a constructive comparison could be made between the phenomenal increase in fuel costs about 1975 and the less sudden, but in various ways more formidable, fuel problem or energy crisis so dramatically resolved by the duke in the eighteenth-century. In the present book I have also emphasised, in the hope that this time it will be more clearly appreciated and understood, the technological significance to the economy of the innovations involved in carrying the first significant English summit-level canal across country, through other men's territory, by the first barge aqueduct in the land, from the new underground canal system at Worsley, towards the hearths and factories of Manchester, Salford, Liverpool, Warrington, Runcorn, Lymm and the wharves and depots between. Despite some able arguments to the contrary, it is not acceptable that the Sankey Navigation, as it was long known, should be endowed with the same technological precedence as the Bridgewater, for although it seems to have been canalised throughout it did not run *across country between watersheds* but followed the set course of a stream. The laurels must therefore still rest with the duke, as his contemporaries clearly judged they should when they designated him 'the father of inland navigation' – the man who could take a waterway wherever a head of water was available.

The significance of the duke's new type of cross-country canals lay in their potential for opening up the geographical hinterland of England to the efficient transportation of bulk goods, and so to economic development. The old river navigations had been, and remained, extremely valuable carriers, but *at that time* they were far more subject to drought in summer and flooding in winter than the canals.

The astonishing increase in trade and population which marked the industrial revolution was caused partly by new machinery processing raw materials ever more rapidly, but transport was equally important. Goods had to be carried, before the railway age, to the mills and away again as finished products, by canals, rivers, road wagons and pack-horses. The roads probably carried the lion's share, but the price of coal was roughly halved wherever the canals went, and, as Dr Fay pointed out, the canals also provided 'a laboratory for experiment in the financing of public utilities, and accustomed a relatively wide range of people in the process to the idea of investing in transferable non-government securities'. Waterway construction also set important examples and precedents in civil engineering know-how and accuracy, but it would be totally erroneous to think as some writers have, of the century and more of canal dominance as a mere *hors d'oeuvre* for the coming railways.

Least understood of all, perhaps, is the part played by family history. In this book we shall see a pattern evolving by which two great houses and their allies vied with each other over the vexed question of the new canal idea. On the one hand stood the powerful Stanleys, Earls of Derby – a lane beside the Irwell still bears their name – whose hereditary links with Salford, Macclesfield, the Wirral and Liverpool virtually obliged them to reinforce the Brookes, the Byroms, and other merchant proponents of, or investors in, the old river navigations. On the other hand stood the innovators of this new canal idea – the duke, Earl Gower, Samuel Egerton, their cousins and allies, including the Gilberts, Wedgwoods, Brindleys and Bills. The North West was therefore still sufficiently conservative to be ranged behind great names and substantial estates for just a little longer.

These properties, some of them considerable, even in an age of landed wealth, were grouped by ties of blood, politics, friendship or common interest, were organised by professional agents and their assistants – clerks, bailiffs and mining engineers, and backed by skilled foremen trained on the traditional system of craft apprenticeship. Some of these estates proved sufficiently adaptable and inventive to launch major enterprises, though the superb engineering of the Bridgewater canal, both above and below ground, running without locks to Runcorn, was clearly something quite unique.

The fact that the duke and his advisers were willing to go further than that and plan a national canal system to link the major ports of Liverpool, Bristol and Hull to the textile centres of Manchester and Salford by 1760 at the latest, does not mean that he was assured or even confident of success in either the Parliamentary or the engineering problems involved. It does mean, though, that the duke was convinced, from his Continental training and observations, that the job could be done, should solve the fuel problem, and would eventually prove highly productive. The outcome of his gamble was the Bridgewater achievement, calling for great courage and tenacity, and so apty summarised in Emerson's epigram: 'An institution,' he wrote, 'is the lengthened shadow of one man.'

For Durand

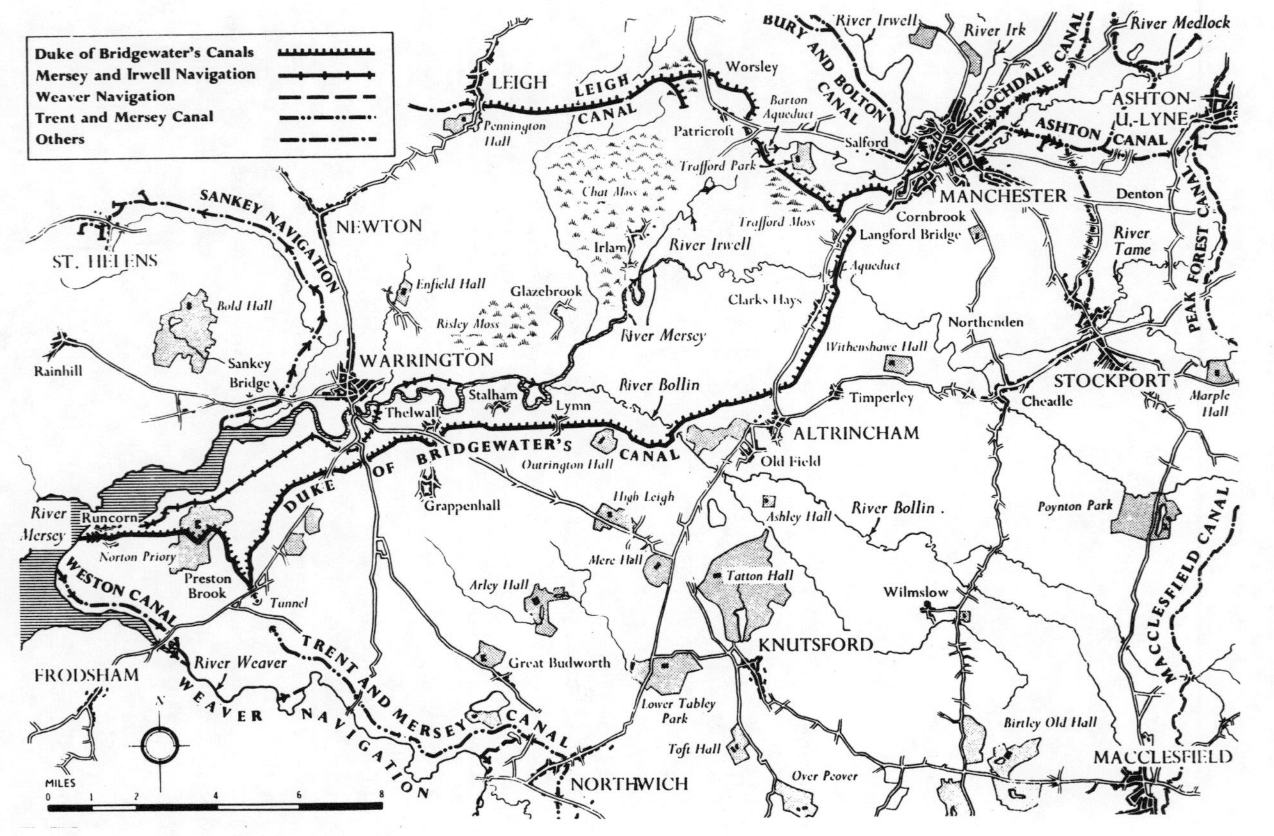

1 The Bridgewater canals and connecting waterways

1

The homeless duke

Times have changed. Canals, with their curved bridges, ancient warehouses and placid waters, drawing the mind away from industrial cities to the heart of the English countryside, now carry more pleasure craft than commercial traffic. In the third Duke of Bridgewater's day it was different – canals then lay at the centre of economic life, creating a transport revolution and assuaging a fuel famine, or energy crisis, even more severe than that of the twentieth century. They influenced the growth of population, reduced widespread unemployment, provided extensive passenger services, aided the agricultural revolution, and served as the main arteries of heavy industry for about a century. No wonder the men who built them, realising something of their potential, became obsessed with the canal idea.

The traveller in the English countryside may still be fortunate enough to catch a glimpse of the vivid, heraldic colours of a working narrow boat vanishing among the shadows, but the future of many canals now lies more with linear parks providing a living industrial archaeology, water supply, conservation and the leisurely world of boating and fishing. Some are being dredged and reconstructed by far-sighted local councils, often aided by voluntary labour, and the Canal Revival has certainly come to stay. Others remain commercially viable, and when we read that 'waterways are much more efficient than roads for freight transport. The number of tonne-miles per gallon is 250 compared with 58 for roads',[1] we may well regret, as fuel grows ever dearer, that successive governments have ignored the repeated advice of commissions and enquiries to bring at least some part of the system fully up to date.

The technologist mainly responsible for introducing the canal idea into England was a practical man with some engineering training[2] who dedicated himself to these two main problems of energy supply and canal construction. Though he suffered from tuberculosis when young, he was single-minded and dedicated, willingly making personal sacrifices to ensure that his waterways would enrich his heritage, relieve poverty, and serve as a model for the nation, though not before he had suffered the risks and ridicule inherent in most new undertakings. This is therefore the life story of a technical innovator whose work has been vindicated by more than two centuries of inland navigation, and as the duke delegated responsibility skilfully, it is also, in some measure, the story of those who served with him – the Gowers, the Egertons, the Gilberts, James Brindley, Wedgwood and Bentley, Robert Lansdale and Henry Tomkinson among them – as well as those who opposed his policies – Lord Strange, the Brookes, Charles Roe, and the Manchester and Liverpool merchants.

The ancestors of the third Duke of Bridgewater stare down from their portraits, handsome and hard-headed, but combining a sense of service with their assiduous pursuit of distinction. The founder of the duke's branch was Thomas, an illegitimate child of Sir Richard Egerton, of Doddleston, in Cheshire, and his servant, Alice Sparke. He chose a legal career, and in 1581 was appointed Solictor General to Elizabeth I. A more than nominal Christian, he treated his friend the fallen Essex with a solicitude rare among rival Tudor courtiers, was created Lord Ellesmere in 1602, and Lord Chancellor when James I succeeded the following year. Thus Alice Sparke's son achieved the first place in the kingdom after the blood royal and the Archbishop of Canterbury.[3]

From his services and through three marriages Thomas Egerton accumulated considerable estates, the principal one being the old abbey and lands of Ashridge, near Berkhamsted in Hertfordshire, which he bought from a private owner in 1604. In 1572 his half-sister Dorothy married Sir Richard Brereton, lord of the manor of Worsley, near Salford, which had a profitable sideline in coal dug from shallow workings on Walkden Moor. The beruffed effigies of the Breretons still gild their tomb in Eccles parish church, which then included Worsley within its boundaries, and as their only child died in infancy this Worsley manor also passed to the Egertons.

A man of some integrity in an age not always noted for honour in high places, Thomas was created Viscount Brackley, and when he died in 1617 his son was granted the earldom of Bridgewater. This title, so very apposite for his distinguished descendant, was taken from the Somersetshire seaport (now spelt Bridgwater) where the family had once held the manor.[4] To celebrate his appointment as Lord President of the Welsh Marches the earl ordered a series of entertainments at Ludlow Castle in September 1634 which included the famous first performance of Milton's *Comus*, in which his son, Lord Brackley, played a leading part – the ruined castle still carries a plaque commemorating the event. When the Civil War shattered the peaceful routine of their lives the earl sympathised with the Royalist cause, but managed to retain most of his estates. The next heir was called John, as was the third earl, who supported William and Mary, and died comparatively young in 1701.

Scroop, fourth Earl of Bridgewater (1681–1745) – father of the canal duke – married first Elizabeth Churchill, the prettiest daughter of the great Duke of Marlborough. Scroop's governing ambition was to climb the last rung of success to the ultimate achievement of a dukedom, but their eldest son, John, died young, and Charles lived only a few months, though their daughter Anne survived to marry Wriothesley Russell, Duke of Bedford. In 1714 Scroop's first wife died, and he married Rachel Russell, sister of that Duke of Bedford whose wife was his own daughter Anne – creating a tangled relationship. In 1720 he was awarded the dukedom of Bridgewater. Scroop had eight children by his second marriage, but all his dynastic hopes were shattered as they were stricken by illness – particularly tuberculosis, an infection which then took a heavy toll of the population and may well have caused his own death in 1745.

Rachel, his widow, intelligent and charming but selfish and self-willed, resisted the consumption and survived her husband by thirty-two years. Scroop had been dead only a few months when, in 1745, she married Richard Lyttelton, who was aged twenty-six, the former duchess being nearly fifty years old, and this ill assorted couple launched forth into a life of gaiety in London – described by Horace Walpole as 'the best-humor'd people in the world'. While Rachel was enjoying an Indian Summer of frivolity her few remaining sickly children were left with paid attendants and generally neglected. Francis was as miserable as any boy of eight who had lost his father, but he survived and was packed off to Dr Pitman's boarding school at Markyate, in Hertforshire, which the poet, William Cowper, had left a few years before.[5]

The children of Scroop's second marriage also died one by one – Charles, his third son, in 1731, aged six; John, who reached his majority and duly succeeded to the title, in 1748; while William and Thomas lived only briefly. The last puling and sickly child was Francis, who was born on 21 May 1736. With the daughters the tubercular scourge acted less virulently. Louisa, born in 1723, survived to marry Earl Gower (1721–1803);[6] Caroline, born in 1724, managed to defeat the family ailment, and lived until 1792. Diana, born in 1731, married Frederic, Lord Baltimore, who from a Stuart grant held extensive lands in North America and the title of Lord Proprietary of the Province of Maryland. Francis later supported him in a border dispute with Messrs Penn & Co., the proprietors of Pennsylvania, but his wife died when only twenty-seven.[7]

It was indeed a depressing toll, for when Scroop also died in 1745 only two of his sons were left alive, and as John, the eldest, seemed superficially more robust, some trouble was taken to groom him for his responsibilities – but he died of smallpox at Eton. At the age of eleven Scroop's seventh son, Francis, a neglected, embittered child, starved of affection and driven almost insane by the persecutions of his young stepfather, succeeded to the highest order of nobility in the land. Mortality must have weighed heavily on his mind as he saw his relations dying one by one, but he soon showed that determined independence which had driven his family to such distinction – though without a father to guide him he found it difficult to make tolerant judgements.

Had Lyttelton himself been wiser and more mature he might have managed to win his stepson's affection and avoid the misery and publicity that followed. He and Rachel had houses in Chelsea and overlooking the Thames at Richmond, and were busy building Cleveland Row, off Piccadilly, but there were bitter and probably violent family quarrels, and Francis either left home of his own accord or was thrown out just after he had gone up to Eton. His official guardians were his mother and his cousin, Samuel Egerton of Tatton (1711–80), an MP for Cheshire from 1754 until his death, who had only one child, a daughter.[8] Francis was attracted by his cousin's integrity and kindness, and though Samuel was inclined to fly into petty rages over trifling matters he remained curiously calm and unperturbed by major mishaps. His brother, Thomas, had worked as an impoverished clerk in Holland for £40 a year, while Samuel had

2 Samuel Egerton, the squire of Tatton Hall, in Cheshire, as painted by Nazzari in Venice (*by courtesy of the Tatton estate*)

made money as a merchant in Venice, as his portrait by Nazzari implies,[9] before inheriting his fine park and estates in Cheshire.

It was on his doorstep that the angry, dishevelled and probably hungry young duke arrived. His guardian could hardly have turned him away, though he was alarmed by this sudden addition to his household, so he took a coach for London to mediate with the Lytteltons.

Rachel was furious, and adamant. A petty woman, she was proud and infatuated, stating categorically that she would not let Francis enter her house again until he had made what she termed 'his proper submission' to her husband. Though Francis was only fifteen years old, he was not going to make a submission to anyone but did go so far as to say that he had a strong affection for his mother, adding as a concession that 'he would never live with or be under the protection of his stepfather, though he was willing to be advised by him'.[10] Those who later bargained with the duke may not always have appreciated the stern school in which he served his apprenticeship.

When asked where he *was* willing to live, Francis said with his uncle, the Duke of Bedford, who had married his half-sister Anne, or with his full sister, Lady Gower, or anywhere with his cousin Samuel Egerton. He did spend the next holiday from school with the Bedfords, where he must have pleaded his cause with skill, for by October 1748 Bedford had decided that there was no justice in his sister dissipating the Bridgewater patrimony in a whirl of London gaiety. He brought a Bill of Complaint in Chancery against the Lytteltons, and even against Samuel Egerton as well, accusing them of misappropriating the personal estate of Francis's father. Bedford insisted that the boy must have the money for his education at Eton which was being angrily withheld by Rachel, and that they had 'intermeddled' with Scroop's estates and 'used the rents and issues and profitts thereof' without any right whatever.[11] Francis, being too young to sue for himself, was described as the plaintiff, with Bedford as his next friend, and the money was promptly made available for his education.

If there is any truth in the maxim that difficult children evolve into able adults it gains support from the progress report written by

Robert Purt, the duke's private tutor, when he arrived at Eton in 1749: 'You, Sir, who are not unacquainted with the disadvantages under which his Grace had labour'd will not expect any account of a great progress made in so short a time. . . . I have procur'd him a place in the school where he will have an opportunity of conversing with manly boys, by which means he will, I hope, become so himself, and wear off any bashfulness which may otherwise stick by him.'[12]

But these beneficial effects were offset by the bullying and misery of the holidays, spent once again with the Lytteltons. Such a bad time did the duke have that it brought an angry letter from Bedford to his sister and her husband: 'I have no pretence to interfere where my nephew lives during the vacations at Eton, and when after he should quit ye school, but . . . if I saw him improperly treated or used at your house, I would as his next relation apply to Mr Egerton, his guardian, to interfere that authority the late Duke of Bridgewater gave him . . .'[13] He even threatened to go to law again. On the other hand neither the Bedfords nor the Gowers were willing to condone his conduct to the extent of offering him a home.

Samuel Egerton of Tatton was kinder, and the copious verbiage of the lawsuit of 1748 clearly shows that the duke was already spending much of his time there. His mining estate lay only a few miles away at Worsley, and he would have learned of his father's hopes of transporting coal along the Irwell to the hearths of Salford and Manchester, and the other plans for a canal, formulated just before Scroop died in 1745.[14] Since the squire of Tatton knew the fine arterial waterways of Italy and Holland, he must often have compared them unfavourably with local navigations, bound to the beds of streams, and subject to flooding in winter and drought in summer. The Mersey and Irwell Act, passed on 17 June 1721,[15] had made it possible for barges to reach Manchester about 1734,[16] though not without difficulty. The Mersey flowed only a few miles from Tatton, and Francis may often have seen its barges riding the flashes, or struggling over shallows, bow-hauled by manpower – a lively scene in that flat, open countryside.

As one of the MPs for Cheshire Samuel Egerton lent his support in Parliament to a projected turnpike road through the duke's Worsley estate,[17] saying that his ward was 'willing and anxious that it should be done', implying that Francis already had a mind of his own in transport matters. In 1754 he also supported a Bill, which failed, for a canal – meaning a waterway independent of river beds, but fed by streams – which would run from Salford to Wigan, drawing water from the Irwell, and carrying coal from the Worsley mines. This had been promoted by a group of Salford and Manchester merchants who were probably mainly proprietors of the Mersey and Irwell Navigation, but objections to the withdrawal of Irwell water on health grounds raised powerful opposition to this early application for a summit-level canal.

Few indeed can be the historical precedents for a duke suing his own mother with the aid of his ducal uncle. Pride on both sides was the chief cause, but Rachel probably needed money, had suffered much in losing so many of her children, and was clearly convinced that it was only a matter of time before the most tiresome of them all succumbed. With clearer understanding of such matters we can see that a child whose mother had married so soon and so incongruously after his father's death would inevitably become fractious and backward if starved of affection. But Francis was no fool. Few people emerge from extreme childhood suffering entirely unscathed; those who survive intact may become endowed with exceptional gifts of endurance and tenacity. For the struggle which lay ahead the young duke was certainly destined to need all the resilience he could muster.

Notes to this chapter are on p. 181

2

The Grand Tour

Bouts of coughing and pains in the chest indicated that the young duke's allotted span in this mortal world might be as brief as his brothers', despite the impressive roll of his titles. But the nation was then ruled by an aristocracy, and, realising that if he did survive he would grow to exercise considerable political power, the Gowers and the Bedfords consulted each other, and decided on the fashionable prescription of sending him off on the Grand Tour. Samuel Egerton was appointed receiver of all the rents and profits of the duke's estates,[1] while Robert Wood agreed to take on the daunting task of tutor and mentor to Francis during his travels. Scholar, traveller and classicist, Wood was the most civilised man available, having recently returned from exploring the ruins of Baalbek, Palmyra and Troy. Author of a popular and distinguished work on his discoveries, as well as a commentary on Homer,[2] he was a polished man of that world of fashion which preferred the precepts of Lord Chesterfield to the preaching of John Wesley, but he was kindly, tolerant and persevering.

For Francis the projected tour of Europe offered a welcome relief from the miseries of a childhood which was making him drink far more than was good for his delicate constitution. With a suitable retinue the distinguished author and the gauche, frequently tipsy and rather unedifying young duke set off for Paris, where Wood tried to persuade his pupil to take an interest in dancing. Francis at first rejected the idea of dallying with anything so effeminate but seized the chance of improving his fencing with the world's greatest masters, and for the first time began to show some enthusiasm for the life around him.

From Paris early in 1753 Wood sent the first confidential report on his charge to the Gowers, assuring them that Francis had survived the rigours of a rough crossing and a very cold journey better than expected. He added grave misgivings that, though he and the duke had ground away at 'some very serious application to reading . . . I can promise very little upon that head and am very much afraid the difficulties are not to be got over'.[3] Since the duke was anxious to see the reports he wrote to the Gowers and Bedfords, Wood thought it wiser to show them to him, because 'his desire of acquiring the esteem of his friends operates more with him than any other motives, and I am glad of catching at anything which may make him exert himself'. Though serveral ladies had gone to the trouble of giving parties in his honour, Wood questioned whether their patience or the duke's would be exhausted first. 'His letters,' he added, 'cost him much trouble, yet, they are entirely his own, tho' generally some errors are pointed out to him which he corrects in his own manner.' In summing up, Wood wrote, 'Such my Lord is the duke's character, a mind as yet untainted, much good nature, and a character very passable but irretrievably neglected as to letters, tho' so equall to his own affairs and the ordinary occurrences of life that I think he may do extreamly [*sic*] well if kept within a proper sphere and in good hands.' Yet he feared that the duke would at any time prefer bad company to a good book. 'However,' he added, little knowing what lay ahead, 'I have taken such precautions against the first that I am under no apprehensions while we stay here.' Wood felt that under these circumstances he need not trouble the Gowers and Bedfords with any further confidential reports unless there were urgent reasons, particularly in view of the duke's anxiety to know what was being written about himself.[4]

Like other deprived children Francis had formed a strong attachment to animals, and especially a pet monkey at Trentham – parting with it had been a desperate wrench, and his letters home were filled with affectionate remembrances to 'ye little monkey' and urgent requests that his sisters should mention it when writing. Under

Wood's kindly guidance his character became less sullen and boorish, and he began to show varied interests in politics, sport and people, including Lord Albemarle, the Earl of Essex and others who enjoyed his sumptuous dinner parties.

From Paris they moved to Lyons, where Francis was to study at the Academy – not exactly a university, but a philosophical school for intellectual training and rigorous physical instruction, in riding and fencing particularly. Wood wrote that as things were going so well he felt that his letters would provide little variety, pointing out that the duke was growing aware of his own shortcomings when he met more polished and civilised people in France, 'and as he knows very well what is expected from his rank, I am sure he has too much spirit to submit to the insignificance and contempt which longer negligence must have brought on'.[5]

Francis's own letters to the Bedfords reveal a genuine affection for the uncle who had supported him at law, in maintaining his estates: 'I propose continuing here a year as your Grace seem'd to think proper when I left England, and shall always be glad in this as in anything else to follow your advice.'[6] Meanwhile Wood was happily trying to obtain what he termed 'apricocks' and peach trees for the gardens at Trentham, when suddenly the peaceful process of educating and civilising Francis was shattered. On 30 May 1755 Wood wrote to both Bedford and Gower tendering his resignation.

My Lord,
I am heartily sorry for the disagreeable surprise it must give the Duke's friends to hear, after what I have so lately wrote in his Grace's favour, that I now beg the Duke of Bedford and your Lordship's leave to return to England – the step I take in so doing is as contrary to my interest as it is necessary to my tranquillity and happiness, and indeed to my character; the Duke and I are on a perfect good footing, tho' he privately acts so contrary to my advice in things very essential to his happiness, nor is there the least interruption of that good nature and complaisance he has always shewn; how long that could be preserved while he continues in the worst company is a question; this seems to me the proper time while we are cool and calm to write to your Lordship; it is not the effect of passion, but of cooll deliberation that I declare I can do the Duke no more service, nor can I, in honour, live with him upon

such a footing . . . the Duke may do well under the care of somebody else and my leaving him may make him exert himself and think properly, tho' my staying probably cannot.[7]

Wood added that another family was anxiously seeking his services, though he felt that he had been treated by the Bedfords and the Gowers in 'the genteelest manner'. If they insisted on his keeping his contract 'I shall expect three hundred pounds a year during my life from the time I leave his Grace . . . and this I must say I shall expect as an act of Justice, not of generosity', but he was willing to stay another six months in that age of slow communications, when a relief could scarcely be found immediately. He added that, though he was not showing this letter to the duke, he would allow the boy to see his next one containing facts, before sending it in a post or two.

On 7 June 1753 Wood wrote as fine a piece of prose as he ever penned, explaining with extreme delicacy the embarrassments to which he had been subjected by his charge's ways:

The more I grew acquainted with his Grace's character (which has been my sole business to studdy since we have been together) the more I grew convinced that the most effectual service I could render him was to give him a taste, if possible, for the company of women of fashion; as I could not hope to make him fall in love with their virtues, I thought it necessary he should hope they had some Vices in which he might find his account; it is not at all difficult in this country to procure him the greatest facility of living with them without any restraint.[8]

Unfortunately the duke, still as shy and introverted as most people who have had a miserable childhood, remained entirely uninterested in the charming lures which his tutor dangled so enticingly before him.

I still watched his inclinations, and found that he was to become the property of the Theatre; this was much better than I expected, and what I imagined might be managed for his advantage; I spoke to him upon the freest footing . . . and only preached up to him some regard to Decency, his purse and health, the only things in his amours I desired to be consulted about. The manner in which he talked to me made me quite easy, when I discovered he had with a good deal of cunning attempted to secure to himself the lowest party's of infamous pleasure; he had, at some expence hired a country house by means of the lowest Scoundrell in ye town who became his companion and Intendant des plaisirs, and who had given the Lady so well her lesson that she refused large sums at the same time that she was every night the property of any Garcon de Boutique who had two louis d'or; upon their expressing their fears of my hearing this the duke told them I might return to England if I disapproved of his behaviour; Champaigne was laid in and very successfully employ'd by the parties concerned, who kept the duke so heated that for two or three days he quite neglected Sir Charles Hotham & my Brother, whom he did not care to trust and seemed to be lost to the lowest pleasures beyond redemption.[9]

What Wood seemed to find unforgivable was not that the duke had shown sufficient enterprise to obtain a mistress, since ethical considerations of that kind did not play a major role in his philosophy, but that Francis had used deception, and had made his own position intolerable by flouting his authority. Moreover there was the awful risk that he might marry the girl, who was patently not the type required. But he acted quickly. First he obtained an assurance from the Governor of the Province that the 'whole Cabal', as he termed it, would be banished from the town whenever he felt it necessary, and then wrote to tender his resignation; but suddenly there came another dramatic change of circumstances.

Whether the duke began to have some scruples of acting with so much ingenuity with regard to me, or suspected I knew something of the matter or was disgusted with the rapacity of his new acquaintances, I can't say, but he prevented my saying or doing anything by coming and acknowledging his follys with tears; I could not then help letting him see a copy of the letter I wrote your Lordship; I was sorry to see him in such distress as it occasion'd and I should think myself answerable for the extremity's he might be drove to if I had not immediately comply'd with his request in writing your Lordship a second letter, which has restored tranquillity; I am afraid of making any remarks on what has happened, & shall leave it to your Lorship to judge what may be hoped for the time to come; the Duke knows nothing of this letter; I shall write again in about a month; I beg my comlim^ts to Lady Louisa and Lady Caroline . . .[10]

While we may look back with a certain wry amusement on these events, they must have been acutely embarrassing for both the boy and his tutor, but they managed to settle down to living and studying together again, though Francis reduced his riding lessons to three a week as the weather grew warmer. The best actor in comedy was chosen to give lessons in French, and by 15 August 1753 Wood was able to add in a letter to the Bedfords, 'His Grace has received visible improvement from the Academy with regard to his person and adress and still more as to health . . . and he is sensible of his wants . . . and preserves good nature enough to be patient of advice when he sees his real interest is intended.' Wood added that Francis had received the Duke of Bedford's letter and was anxious to deserve 'the character he has acquired of a good Correspondent'. The spirit may have been willing enough, but this was not a reputation the duke retained for long.

The summer vacation found them enjoying a picturesque tour through Savoy and Switzerland, though they were back at Lyons by 16 September 1753, when Francis wrote to the Duke of Bedford worrying about a suitable colleague for Leveson at his pocket borough of Brackley. He chose Mark Dickinson, an alderman of the City of London, who later became a leading chairman of canal committees in the Commons, and was of great value in supporting the duke's new projects.

The Canal du Midi

When they returned to Lyons Francis, fearing a further dose of the academic grind, made a suggestion which must have filled his tutor with a good deal of apprehension: he wanted to visit the docks, locks and general works of the famous Languedoc canal in southern France, already something of a tourist attraction for foreigners. Still hoping to survive to inherit the Worsley estate, he may have recalled descriptions by Samuel Egerton and Earl Gower of arterial waterways in France, Italy and Holland, and he wanted to see how the experts had done it. Wood sadly but wisely accepted that the duke's bent was more practical than scholarly, and wrote tactfully to Bedford to gain permission, assuring him that they would not 'look upon it as a party of pleasure' but would study the unique antiquities there, and 'go into the best French company'.[11] By 9 December 1753 approval for this strange interest in engineering had come through from Bedford in London, who agreed, equally tactfully, that keeping good company was the best education and an excellent way of dispelling that shyness which he held to be the traditional bane of English youth.

The docks and canals of Languedoc, though inestimably superior to the Irwell's muddy stream and tiny quays, were the last things likely to appeal to Wood's refined and aesthetic nature, and we can almost feel the relief from boredom and the discomforts of coaching when he wrote to the Bedfords on 4 February 1754 from Aix-en-Provence, 'we have been now a month from Lyons, the Duke of Bridgewater has satisfied his curiosity with regard to Marseilles, Toulon and this place, we shall probably take the road of Arles and Nismes to Montpelier where the Duke de Richelieu is arrived to assemble the Estates of the Provinces'. From there they travelled on to Sète, the Mediterranean port of the Languedoc canal, now called the Canal du Midi.

Their rumbling, bumping coach bore them along the course of this waterway, which links the Atlantic at Bordeaux by a canal 150 miles long, starting at Toulouse, with the Mediterranean at Sète, on the Étang de Thau, some 320 miles away. To join these oceans was an ancient idea, but scarcely feasible in a land with so little rainfall, until an understanding of mitre locks had spread to France, for these saved much wastage of water. (The mitre-gated locks, meeting at a wedge-shaped point, held water, enabling boats to be raised or lowered by controlling sluices. Still in use, they superseded the earlier flash locks). The engineering genius mainly responsible for this achievement was Baron P. P. Riquet de Bonrepos (1604–80), who had encountered appalling difficulties constructing the greatest feat of engineering in Europe between Roman times and the nineteenth century.[12] The Languedoc is carried for considerable distances on embankments and has many aqueducts, tunnels and cuttings, and the duke would have noted that, although it took only nineteen years to build, it employed an army of 8,000 labourers. The French canals were free from tolls once they had paid off their construction costs and as this waterway saved the long haul around Spain for sailing ships, it was, by 1724, some forty-three years after it had been built, becoming a valuable strategic and commercial asset.

Francis was deeply impressed by Riquet's struggle against natural hazards and new techniques of engineering, and noted that, although it followed the bed of the Garonne for some distance, on the canal section between Agde and Toulouse it was completely independent of rivers, the summit being supplied by a vast lake and feeders from mountain streams. In England, which was centuries behind the Continent, there were no summit canals. There the navigations obtained their water supplies by clinging to the beds of rivers, so subject to flooding in winter and drought in summer that trade was being strangled. Only on the Newry Canal, in the north of Ireland, had the canal idea taken root, and that excellent waterway was probably planned by Huguenot refugees from the religious persecutions in France.[13] It therefore follows that the Continent was centuries ahead of England in this technology, while the wealth which encouraged the French to undertake the Napoleonic wars stemmed partly from this highly effective and long-established transport system.[14]

At Montpelier the Duc de Richelieu, who was busy with political affairs in the province, had paid Francis the courtesy of returning his call, and then they had set out through Sète, Carcassonne and Toulouse for Bordeaux, a journey taking almost two weeks. From Bordeaux fine weather and good roads tempted them to include the Loire in their itinerary – still partly a working navigation but already silting up – and by 18 May 1754 Wood was bravely claiming that his pupil had seen most of the French provinces and met the best company in them, though what he privately thought of docks and locks by that time was probably unprintable. Francis did indeed show some interest in people like Richelieu, but what really fired his imagination was the canal works he had seen. The canal idea took root, and he started attending a course of what Wood termed 'Experimentall Philosophy,' which included science and engineering.[15]

Francis also began to learn Italian in preparation for his visit to Rome, and startled his tutor by taking up music – 'with what success,' wrote Wood with a twinkle in his eye, 'I can't say, as he don't yet venture to perform in publick.' He added that, since the duke was drinking less, he hoped that he had 'got the better of the passion', though the lavish dinner parties he was giving instead to friends like Lord Essex meant that they had so overspent their £2,000 a year allowance that he was obliged to lend Francis £400 – a windfall from the French translation of his book on the Palmyra antiquities. Even these princely entertainments ought, he felt, to be indulged as long as they kept him from more dubious vices. 'The duke's good nature should be added to the account. I flatter myself upon the whole that tho' all sort of improvement is irretrievably given up, he may pass through life very well under your Grace's direction – ' a prophecy memorable for its modesty.

On 24 September 1754 Francis wrote a worried letter to Bedford confirming the alarming state of their finances, and for once answered his uncle promptly: 'My Dear Lord I write this to return you thanks for yours of the 20 first and to desire your Grace to be so good as to speak in my favour to Mr Egerton to get my allowance increased, for upon noting up our accounts, upon leaving this, we were four or five hundred pounds behind hand – our first quarter, and if it had not been for Mr Wood's having had that sum by him we must have stuck in the Mud. And I can assure your Grace without any extravagancy . . .'[16]

By 10 October 1754 they reached Nice and tried to catch a galley to Italy, but it was so crowded that they were forced to wait, which proved a blessing. Francis had stayed out too late on the water at a party given to honour a visiting prince, and had a recurrence of his symptoms, with a cough, pain in the chest and a fever. The Governor, an old friend of Wood's family, put them up in his own house, whence the tutor wrote cheerfully that the duke's relapse had 'almost given way to two bleedings'.[17]

Francis had tried to learn some Italian at Lyons, and when he reached Rome they called on most of the cardinals and ambassadors, and toured the sites. Wood mentioned in his next report that the duke 'would have waited upon His Holiness along with me, but chose to defer it till he has got a little more Italian, which will be very soon as he goes on'. There were some twenty young Englishmen in Rome, so they led a gay life – too gay, as it proved.

While Rome did much to civilise Francis, the main achievement of that winter was revealed in Wood's report of 9 January 1755, in which he dismissed any fears that the duke might become a spendthrift and said that, although he had decided against rebuilding Ashridge in the classical style, he was determined to devote one room entirely to pictures by hanging them in the crowded fashion of that day, and they would be by the best masters. As the duke had spent all his money again, Wood had to lend him another £400, and so the magnificent Bridgewater collection of paintings, like his other achievement, canal building, began, propitiously enough, with a substantial loan.[18]

The gaiety proved too much. Francis suddenly went down with a further serious relapse, and for weeks on end Wood struggled to keep him alive, aided by Brand, a friend of the Bedfords, who happened to be in Rome. Slowly Francis regained a little strength. By 24 May 1755 Wood wrote that he was a good deal better, while his illness had persuaded him of the need for temperance – 'In short he runs the greatest risk of a consumption upon the least excess, which I think he will guard against, as the physician has been very explicit with him' – so that Francis was obliged to cut out his bouts of heavy drinking.[19]

Recovery took longer than expected – by 12 July the duke was still on a diet, but his tutor was writing that he had as good a prospect of health as anyone as long as he behaved in such a way as to deserve it, and, though the good resolutions to moderate drinking might last a while, all would depend on the company that he kept in future; yet many of his friends in Rome still feared that he would never see England again. By 8 September 1755 Wood was writing in a more cheerful vein from Aix-en-Provence, on their way home, that Francis had 'passed through the danger of setting out in the dog days and of lying some nights at sea in an open boat, without the least bad consequences to his health', and indeed fresh air was probably far better for him than gourmandising in the stuffy salons of Rome, while he added that he had greatly reduced his addiction to 'wines and strong collations'.[20]

Very occasionally, in cases of severe tuberculosis, the main mass of the infection dried up, and was enveloped in a calcareous tissue; the infection became, as it were, walled in, and was cured, or remained quiescent. It is conceivable that this is what happened to the Duke of

Bridgewater. Certainly he returned a different person from the once gauche and hesitant schoolboy: he had received, during two years, the devoted and tolerant care of one of the most polished and widely travelled men of that age; he had met many of the most civilised and brilliant people in Europe; he had mingled with eminent painters in Italy, and had laid the foundations of his famous Bridgewater collection. Francis was a practical creature, while Wood was a scholar and dilettante. That they got on so well is a tribute to them both, but writers who fancy that the duke was an illiterate simpleton could hardly be more wrong: he had shown considerable diligence in learning fencing, engineering and Italian, and could read French well; he had studied the waterways and docks minutely, while his lapses at Lyons had taught him just what Wood wanted him to learn – the virtue of discretion!

And so they came home to England – Wood to fill a vacant seat at Brackley with distinction, and to hold posts in successive governments, and Francis to less serious pursuits centring round the gambling tables, race tracks and various ladies' parlours. Yet there remains a relic of their tour. While they were in Rome the duke persuaded Wood to sit for his portrait to Anton Raphael Mengs, a Bohemian who was court painter to Charles III of Spain. This picture, which was treasured by Francis, still belongs to his sister's descendants.

Notes to this chapter are on p. 182

3

Of soughs and ale

Since Francis was still only nineteen when he returned from the Continent, he had two more years to fill before Samuel Egerton handed over to him the broad estates which his family had accumulated. Scroop had found that his impressive titles weighed heavily on his finances, and did not feel justified in paying the few hundred pounds which would have carried the dukedom to collateral branches, so his last son's immediate duty was to find a suitable wife, to ensure that his own branch of the family would continue. This fitted in tolerably well with his keen determination to enjoy himself and lead a fashionable life at Bath, London and Newmarket.[1]

The world in which the young duke took his place was divided between two major powers, France and England, competing for a supremacy once held by Spain. While Francis and his tutor were on the grand tour Clive was conducting his campaigns against the French in India, and there was trouble enough in North America. Tension built up within and outside Europe, and in 1756 boiled over into the Seven Years War. A threat of invasion from France was countered by importing German troops to assist the English, who were, as usual, doing their utmost to lose the first few battles, but when the elder Pitt came to power in 1757 the tide gradually began to turn in favour of the Anglo-Prussian alliance.

The home front was equally insecure; it was only ten years since the Jacobite rising of 1745, and though the capable and cynical Walpole was dead his philosophy permeated the land, branding the age with the justifiable soubriquet 'Soul extinct – stomach well alive'. England was ruled by a small oligarchy of landowners who derived their power from vast estates which were run for them by capable professional land agents. The population of about six million[2] was still mainly employed in agricultural or allied occupations, but industry and commerce were thriving, although still much hampered by muddy roads. The shallow river navigations also carried a great deal more goods than is generally realised.

Francis soon showed that once he had been released from even the tolerant guidance of his tutor he was determined to lead the life of a man about town. In London he gambled at White's, and in the country began his turf career seriously with a house and stables at Newmarket, which he maintained until 1770, when his debts for canal construction had grown to astronomical proportions.[3] He was not too tall – only 5 ft 9 in. – and though still very delicate he worked off most of the ill effects of London life by racing and exercising his own horses.[4] He reduced his weight so well that one windy day a wager was taken that he would be blown out of the saddle before reaching the finishing post. A fundamental change had come over his character. In France he had been reluctant to take riding lessons: bearing the world a justifiable grudge, he was at first reluctant to do anything but drink. Now he relished his days in the saddle; the most famous of his horses was the Cullen Arabian, bred in 1757, but studmen also revere the records of Star and Astridge Ball.[5] His turf companions, not a particularly savoury collection, included the Duke of Cumberland, second son of George II, and the fourth Duke of Queensbery, 'a rare judge of horseflesh, to emphasise the most attractive of his largely questionable accomplishments'.[6] Blue silk and silver were the duke's racing colours, but by 1764 he was auctioning many hunters from his Ashridge stable 'at the sign of the Robbin Hood on Little Gaddesden Hill Herts' to meet the rising cost of canal construction.[7]

The broken engagement

When Francis was only sixteen the *Manchester Mercury* had forecast his marriage to Jane Revell, the heiress of a wealthy government contractor who was MP for Dover.[8] Her estates in four different counties and capital of £250,000 were sufficient to revive this rumour when the duke returned to England in 1755, but he dithered so long that she grew tired of waiting and eloped over the border with George Warren, an impoverished Cheshire squire, who was later to join the ranks of the Old Navigators, and others who rigidly opposed the Gower and Egerton canal interests.[9] The duke's next encounter was with an officer's widow whom he helped to a second marriage with a grant of £500, but these were mere preliminary forays for the unhappy affair which followed. In 1751 Elizabeth Gunning and her sister Maria were brought out into the world of fashion by their mother Bridget, the daughter of Viscount Mayo and the wife of John Gunning, an impecunious Irish squire from Castle Coote, County Roscommon. When the time came for them to be presented to the

3 Elizabeth Gunning, 1733–90, who was engaged to be married to the duke

Lord Lieutenant at Dublin Castle they were too poor to buy dresses, so Sheridan, out of the kindness of his heart, lent them two magnificent gowns from the theatre. At the ball which followed they became the rage of the town, with their beauty and vivacity, and, not content with conquering Dublin, moved on to London, where they nearly had to work as actresses to earn a living. But such loveliness could not long go unrewarded. They were mobbed at the door when they tried to go out, theatres were packed when it was rumoured that they would be there, and like enchanted children in a fairy tale the doors of the palaces of London opened before them.

In this era of 'teenage idolatry' the homage to the Gunning sisters may not seem eccentric, but at the time it was the subject of much comment. Royal instructions were ordained by George II that when they strolled down the Mall on Sundays a platoon of the Guards should keep their way clear and protect them from their fans. But all the best fairy stories end in marriage – Maria's, the eldest, to the sixth Earl of Coventry, and Elizabeth's to the Duke of Hamilton, who fell head over heels in love with her at a masquerade. Walpole relates how Hamilton found himself alone with Elizabeth, and sent for a priest, who very properly refused to perform the ceremony without a licence or a ring, but despite his objections they were wedded at half past twelve at night with a ring snatched from a four-poster bed.

Hamilton did not long survive to enjoy the prize which his impetuosity had won him; the widowed Elizabeth was now not only beautiful but also rich, and though he was three years younger, Francis Egerton fell deeply in love with her. He proposed, and to his surprise, since he had not yet outrun the shyness which had so long dogged his character, he was accepted. Preparations were soon being made for the marriage, when suddenly the dark shadow of scandal fell over the Gunning sisters. London society was almost as openly salacious as it had been in Charles II's days, but there were still a few things from which it shied away, and when Maria had an *affaire* with Lord Bolingbroke under the nose of her husband it became the talk of the town. By and large, people were not amused.

The whole sad business was related in a wise and understanding letter from Mrs Wood to the duke's sisters, Lady Louisa Gower and

Lady Caroline, who were still living together at Trentham.[10] Though Mrs Wood had returned fatigued to dinner, she could not help writing at once to let their ladyships know that Francis's affair with Elizabeth, Duchess of Hamilton was most probably entirely off. Willing tongues had carried the scandalous news of Maria's doings to his ears, so he had written an extremely careful and courteous letter to Elizabeth pointing out that he considered his attention to her sister as a compliment to her Grace, whose prudence and friendly assistance he so greatly valued. Although this letter was, Mrs Wood explained, 'conceived in very cautious and respectfull terms, her answer was unexpectedly high and decisive (especially after kinder and more affectionate ones that your Ladyship saw) she insisted upon an absolute priviledge of not being advised about *any Company she should think proper to keep*, terms without which she desired to be entirely off, with more in the same style'.[11] The duke's reply expressed the deepest mortification and concern – he only hoped that her letter had been written in a passing passion, insisting that he had never for a moment intended to sever 'that laudable sisterly intercourse' which was so clearly a virtue, but on the other hand he felt that he could hardly be expected to be so disingenuous as to approve of Maria's recent conduct. (Mrs Wood added that all this had been written far more tactfully, and in greater detail than she had expressed it.) The duke desperately hoped that Elizabeth would not consider his hesitations about her sister as any mark of his diffidence towards herself, added warm and tender expressions of his regard, begged that she would reconsider her view, and agreed that she should keep whatever company she pleased, though there was a decorum which he knew she respected, while he certainly did not mean to imply that her sister was to be avoided.[12]

Poor Francis must have struggled painfully with the delicate phrasing of these letters, and could scarcely have gone further or behaved more abjectly in propitiating that proud beauty, but he was unaware that his heart was already being weighed unevenly in the balance against handsome Colonel Campbell, the future Duke of Argyll, a man of the world, far older and more experienced in the arts of love and war. Elizabeth's final answer, which the duke long awaited with misery and impatience, was in exactly the same vein as her earlier one, so he took his sorrows to a wise man called Mr Bass, who thoroughly approved of his conduct, and after this, as Mrs Wood put it, the duke 'bid adieu to Scotland', though she added that he suffered immensely.[13] She continued, 'I am glad his G. takes so much to Mr Bass, who I hope will go to Ashridge with him, I never knew his Grace behave with more good sense, calmness and discretion. I hope what is done will have your approbation and say nothing of Ld. Gower who I hear is at Bedford.'[14] Clearly the Woods, who knew so much of the duke's earlier and even less appropriate *amours*, had been his kindly confidants throughout.

By January 1759 Horace Walpole was writing to a friend about the match between Elizabeth and Colonel Campbell, claiming that everybody liked it except the Duke of Bridgewater and Lord Conway, and insisting that he himself would never dare to marry either of the Gunning sisters for fear of being 'shuffled out of the world prematurely to make room for the rest of their adventures'.[15] Since Maria had managed to marry an earl, and her sister had achieved a hat trick of dukes, their rise from rags to riches gave the Irish beggars a popular saying when offered largesse: 'God rest your honour, and may the luck of the Gunnings attend you!' Elizabeth's subsequent record was not discreditable either, as she ended her days as the mother of four dukes, two of Hamilton and two of Argyll.

For Francis Egerton this traumatic disappointment almost set the final seal on his series of unhappy relationships with women – his mother had spurned his affection and finally rejected him, his brief escapade with the actress at Lyons had created untold misery and embarrassment, and finally he had been obliged to break off his engagement to one of the loveliest girls in the land.[16] A less persevering man might have reacted less wisely, but these misfortunes spurred him on to prove that, even if he might be the last of his branch, he was capable of serious work, and could and would make his voice heard for a moment in the silence of eternity. Though he had not altogether neglected his estates, he had not visited them any too often. He now turned his back on the dicing tables of London and Bath, and the touts and rakes of what was cynically termed King

Arthur's Square Table came to consider him a lost man. They failed to appreciate that he was only rejecting their brand of gambling in order to engage in a far more momentuous wager.

The duke's resources

The Egertons had shown a continuous skill in business matters. Their titles were grounded on ability and wealth drawn from the rich acres which succeeding generations had so carefully accumulated. Though Scroop had acquired a quiverful of mortgages, he was still able to leave his last son substantial estates in twelve different counties. The family home and headquarters was the old abbey of Ashridge, near Tring in Hertfordshire, which was managed by William Tyler, and extended into Bedford, Buckinghamshire and Middlesex.[17] This produced an annual net income of about £4,000 – a sum which fluctuated by as much as £400 in a single year, even though it relied heavily on farm rents, while it was burdened with an annuity for Caroline, the duke's unmarried sister.[18] He also held substantial estates at Whitchurch in Shropshire and around Ellesmere, which were probably managed jointly with the Gower holdings. The pocket borough and manors at Brackley in Northamptonshire were valuable, while the estate at Bracken, Norton in the Clay and Caldwell in Yorkshire produced about £1,000 a year until Francis's loans to tenants for improvements greatly increased its rental;[19] even further afield lay the Durham, Westmorland and Suffolk properties. All his twelve estates, together with Worsley, produced an average annual income of only £30,000 a year gross,[20] but in addition Scroop may have left his son some £50,000 in government bonds.[21] The conclusion must be that although Francis Egerton was, in Forsyte parlance, a warm man, his property was widely dispersed, somewhat encumbered, and costly to administer. By eighteenth-century ducal standards he was not considered wealthy, and had heavy commitments, including salaries, capital investment and maintenance. Clearly there were sound reasons for dedicating his life to improving his estates rather than following the other family tradition of politics, while time was to prove that his resources were scarcely sufficient to enable him to carry through his grandiose canal plans without risking his inheritance.

The estate which was destined to become most valuable his father had tied up in an entail which remained in force during Francis's lifetime, preventing him from selling or mortgaging any of the land.[22] This was centred on the manor of Worsley, some seven miles from Manchester. Coal had been mined there in medieval times for local use (the first surviving document being dated 1376), and a lease of 1575 allowed the colliers 'To come with horses and cartes, carryages and workmen, to dygg and carry awaye all such colles as shall be found growin within the desmesnes of the Peele of Hulton . . . and that without any fraude or guyle'.[23] The word 'growing' used in this context is curious: peat was long thought of as a growth; a similar superstition also applied to coal and even to stone. By 1724, and probably well before that time, the Egertons were employing men regularly in mining their coal for wages varying from 10*d* to a shilling a day.[24]

Over the centuries most of the seams near the surface had been worked out, and well before Scroop's time (1681–1745) the shafts were delving further into the moor. These pits were still comparatively shallow, for the usual method of draining them was 'horse-ginning'. A horse walked round and round a winch, drawing buckets of water to the surface, and when these were emptied the contents percolated slowly back to the level from which they had recently emerged.

John Massey was Scroop's agent at Worsley; his stewardship began about 1721 and lasted until 1745. Prices were still so stable that his salary of £50 a year never altered.[25] The net profit (meaning free from all charges and encumbrances) from the Hulton mine in 1725 was some £367, while Worsley mine brought in only about £301,[26] and the money was despatched every year to Sir Francis Child, Scroop's banker in London.

On the Worsley estate Scroop had been faced with two major problems: to get his coal from flooded mines under Walkden Moor, and to transport it to the towns. By 1743 the pits were nearly drowned out, and the combined net profit had slumped to a mere £122, far less than the farm rents, which came to over £512. As these shallow diggings became unworkable, desperate efforts were made to find new ones – Massey even tipped a shilling to 'Old John Collier for letting us see where the coals Rank [lie] on Walken Moor'.[27] Paying one of the gaffers a shilling for his trouble seems a somewhat incongruous method of finding a new pit. A host of entries in the accounts illustrate the desperate battle waged against the water level and the vast expense entailed: 'Pd Oliver Darbeshire, James Siddall, for lading water into ye buckets in ye bottom of ye same pit at 1/- per day for 6 days . . . 12/-'.[28]

There was one other way of relieving the drowned mines, and that was by means of a sough. A sough is simply a drainage tunnel drilled into the side of a hill at a slight angle until it reaches the coal seam and releases the water. Soughs were not new; their function in the Worsley area was described at the Dixon Green Colliery, where the man holding the mining lease was empowered in 1647 to make 'mynes or soughes for the draining of the said coal or Cannell dry'.[29] There was another at Farnworth, but the main one at Worsley was so ancient that no record remains of when it was first begun, though it ran across Walkden Moor from south to north, on a higher level than the later sough at the Delph. (See Appendix A.) Scroop and Massey may well have planned to build locks right up to the sough mouth, for on 13 September 1735 they paid for a dinner and ale to celebrate the surveying (then called levelling) of the ground between 'old sough mouth Worsley Mill and from thence to Middlewood', a plan which would have enabled them to bring boats as close as possible to the coal face.[30] Indeed, boats were probably being used in the sough, for an entry by Massey in 1743 records that coal was definitely being brought from the mine via this adit.

The trouble was that this old sough ran far too close to the surface, and cost the estate thousands of pounds in maintenance over the years. From 1724 onwards there are endless entries in Massey's accounts for repairing it, in one of which the roof is referred to as 'part stone, part timber which fell in'.[31] No sooner had there been a fall than the miners were sent in to release it, and unleash the dammed-up water on themselves; their courage was the foundation of a magnificent tradition, for their labours were lit only by flickering mutton-fat candles made by their wives – the first satisfactory safety lamp was invented by George Stephenson of locomotive fame in 1815, and the Davy lamp in the same year. The sough work, being wet and particularly unpleasant, always earned them an extra allowance of ale, as is shown in Massey's entry for 1726: 'For Drinke for 27 men that was in the bottom of the sough all at wontime. The sough was decade and the water was stoped and there was no other way to lose it but thorow the narrow walls and the walls being so Streate that the fallen stone and Earth could not be drawn thorow but the men was forced to lye on their bales in the water one behind another and remove the stuf one to another.'[32]

These endless difficulties in the main Worsley sough brought expenditure close to receipts, but if the mines were to go on working the springs under the moor had to be released. Quite apart from labour and timber there were innumerable burdensome expenses for the estate caused by flooding, including coal for drying the men's clothes, special issues of flannel shirts, and ale in vast quantities.[33] Even when the sough was in fair working order Scroop and Massey were unable to increase the mining profits much, for although the seams became dry enough to work they were still faced with the graver problem of transporting coal to nearby towns. This and much more about his Worsley estate the duke could have learned from his earliest childhood at Tatton, and when he had seen the French waterways and studied engineering at Lyons he became convinced that it would be perfectly feasible to introduce the canal idea into England also.

Notes to this chapter are on p. 182

4

The canal idea

The most important factor in understanding the transport revolution introduced by the young duke and his advisers is to appreciate the basic difference between a navigation and a canal: a navigation uses a river, with artificial cuts alongside the shallows, draws its water from it, and must therefore follow the fall of the stream. Canals are quite different: they usually draw their water from feeders or reservoirs, and can therefore go wherever an adequate summit level of water can be obtained. The so-called 'still water' canal is something of a misnomer, because canals are never still unless they are dead – for water is always moving from their summits to the sea – but they were a major technological advance on those old navigations which stud the parliamentary records before the duke's day and had been in operation for centuries.[1] The Thames is a good example of a navigation, while the Grand Union is a typical canal.

The two young men who pondered over their maps at Worsley Old Hall in 1757, in a room which can still be seen, were not the first canal builders in England. Part of the Fossdyke dug by the Romans at Lincoln is still in use, John Trew had built a small boat canal near Exeter between 1562 and 1566, and in Lancashire in 1755 an Act was obtained to build the Sankey Navigation or St Helens Canal – so called because it drew water only from the upper reaches and was in all other respects a canal, though not a summit-level one.[2] Francis and John Gilbert certainly learned a great deal from Berry's work on the Sankey, and labour was occasionally exchanged between the two enterprises.

Although there had been much improvement in some navigations, the little Mersey and Irwell, which ran close to Tatton and is now largely lost in the Ship Canal, typified the shortcomings of using small rivers for transport. It silted up regularly, was often short of water, and was studded with snags and shoals. Unhappy merchants might find their goods delayed for weeks by such natural hazards. Even before Scroop's interest in waterways began, and he had travelled in Holland and Italy, there had been proposals by Thomas Steers and others to make the Mersey and Irwell more fully navigable, but the financial climate was not encouraging.[3] At last an Act was passed on 17 June 1721,[4] authorising locks, artificial cuttings and a towpath from Warrington to Manchester, with a toll of 3*s* 4*d* for the use of all or any part of this waterway, except on marl and manure – a clause generally attached to waterway Acts because contemporary ethical thinking insisted that benevolence and agricultural and social improvement should be an obligation of any undertaking capable of reaping private profit.

Slow indeed was the progress in building the Mersey and Irwell. Work did not begin for four years, since the failure of the South Sea Bubble in 1720 was still fresh in the minds of the investing public, but it had reached Barton lock by 1724, when the Worsley estate charged the promoters £42 for a rood of stone for lock walls and surrounds.[5] It was 1734 before all the waterway could be said to be open after a fashion,[6] but even then it made no money for its promoters, who suffered frequent calls on their capital. These investors had a vision of the potential trade between Liverpool, Warrington and Manchester, but it was a clear case of bread today and jam at a very indeterminate future date.

One of the objects of the promoters of this little waterway was to supply coal to Salford and Manchester. There had been plans to bring it from the Douglas navigation, farther north in the Lancashire coalfield, but this did not materialise, so the proprietors approached Scroop Egerton, knowing his estate and others in the hinterland to be rich in coal, and drew up a draft agreement with him to make Worsley Brook navigable to carry to the growing market for fuel in

Manchester.[7] The Egerton corn mills at Worsley were, by this draft agreement, to be moved higher upstream to provide more water for boats running up to Chat Moss near the Old Hall, from which wagon railroads were to radiate to other mines on Walkden Moor.[8] Water supply was important, and the proprietors of the old navigation insisted on using soughs and streams on the Worsley estate, which would have discharged into their main line and relieved the shortage there,[9] while the parties also bound each other by a bond of £10,000 each for the fulfilment of their obligations. Here, then, was a large mining estate seriously considering co-operating with other mines to grant access to a navigation to transport coal, but there is no indication that anything was signed or sealed – if it had been the Mersey and Irwell would have owed the Egertons a substantial indemnity for failing to build this branch waterway. An idea of the already considerable output of the surrounding mines can be gained from the stipulation that between five and ten thousand baskets of coal a week were agreed as feasible.[10]

An Act was passed in 1737[11] to make Worsley Brook navigable for two miles from near the mining area to the Irwell, and this narrow stream was surveyed by Thomas Steers, but possibly because of his exalted rank Scroop Egerton stood a little apart from this – he was named as the first of the commissioners appointed to arbitrate between promoters and other interested parties, and is therefore unlikely to have been a promoter himself.[12] Nonetheless the Egertons clearly stood to gain from cheaper coal freights and maintained that there was an unfulfilled obligation on the promoters to carry for them and their neighbours on the brook, for this was the foundation stone of Francis's first petition to Parliament for an Act to construct the first significant canal system in England.[13]

It was probably the cost of lock construction at a time when the main line was not yet paying that stymied the branch line, and the Irwell made no financial breakthrough until about 1753.[14] By 1758 Francis Reynolds of Strangeways was writing to Edward Chetham that 'The Duke of Bridgewater is come into the country to visit his estate of Worsley, and does me the honour to take up his quarters with me . . . his Grace has found so large a Mine of Coal, for which he has so small a consumption, that he is inclinable to make a water road from Worsley Mill to Salford, at his own expence, by which means he will be able to supply Manchester at a much cheaper rate'.[15] It is now clear that the duke's determination to put the canal idea into practice was not influenced by the end of his relations with the Gunning sisters, since his break with Elizabeth came in November 1758[16] and on the 25th of that month his petition, which must have taken much time and thought to prepare, was first presented to Parliament.[17]

So much has been written about James Brindley's contribution to the Bridgewater Canal that the achievement of the duke and his Resident Engineer, as we would now call John Gilbert,[18] has been gravely underrated. There is no question that Brindley was a great engineer whose skill, enthusiasm and inventiveness carried out the canal idea across England.[19] But there is no evidence that he was employed at Worsley until well after the work on the sough and on the Salford Canal had begun. Nor has the part played by the duke's advisers been clearly understood. The first of these was his brother-in-law, Lord Gower (1721–1803), a man endowed with considerable charm, intelligence and personality, whose land agent, Thomas

4 Granville Leveson Gower, first Marquis of Stafford, 1721–1803

Gilbert, was also General Agent for the duke.[20] This means that these two considerable estates, powerful not only financially but also politically, were run in close alliance with each other and with that of Samuel Egerton of Tatton and Ordsall Hall in Salford.[21] In that age of landed wealth such a combination formed a powerful hegemony.

Meanwhile, Bedford and Gower bore in mind the wise words of the duke's old tutor, that if that ebullient young gentleman was kept in proper hands he would do well enough under their guidance. There could scarcely be a more useful service to the community than reducing the price of fuel in Salford and Manchester, with the added advantage that a canal should eventually pay well, as successive generations of Egertons had realised. Although the duke knew his own mind he did not have a closed mind, and, being open to advice, would have welcomed a course which satisfied his practical inclinations; the canal idea must also have been a recurrent topic of conversation during his holidays from school at Tatton and Trentham.

Since Francis could scarcely remember his father, he modelled his character to some extent on Gower's, and even though his sister Louisa Egerton died in 1761 they remained lifelong friends, despite the earl's seniority in age by some fifteen years. After his own father's death in 1754 most of Gower's energy was dedicated to politics. He became Lord Privy Seal, Master of the Horse and Lord Chamberlain, but he managed to devote time to encouraging canals like the Trent and Mersey, cutting his Donnington Wood canal about 1765, and to drilling mines and navigable soughs on his Lilleshall estate near Coalbrookdale in Shropshire. In 1783, on the fall of Lord North's administration, he rejected the offer to form a Ministry, otherwise his name and reputation as an orator might be better known in the annals of constitutional history. He was a very wealthy man, wielding considerable political power, and a member of that Whig party nicknamed the 'Bloomsbury Gang' which was led by his brother-in-law, the fourth Duke of Bedford. His last wife, Susannah Stewart, second daughter of Lord Galloway, wrote a glowing tribute: 'I believe his good principles, his justice, his truth, his honest upright heart make peace within which diffuses itself upon his countenance, and exhilirates himself as well as those who have the happiness to live with him.' Such a personality could not fail to influence a young man in Francis's position, but it is hard to assess just how much the canal idea owes to him, even though he had engaged Brindley to survey the Trent and Mersey in 1758, at a time when Francis was also working on a canal and other improvements at Worsley. It may be that he was not quite so much of a gambler as his young friend and relation, or so willing to make the first throw in an enterprise fraught with so many uncertainties.[22]

The Gilberts

The next most influential figure in the duke's life was probably Thomas Gilbert (1720–98), who was his General Land Agent, and also chief legal adviser for Francis and Earl Gower. Historians have

5 Thomas Gilbert, MP, JP, 1720–98, General Agent for six of the Bridgewater estates, and Poor Law reformer

often been puzzled that the duke left so few letters behind and, jumping to conclusions, have erroneously assumed that he was an illiterate simpleton. Nothing could be further from the truth. Francis was generally far too busy to write his own letters: he simply told Thomas what his policy was and then checked the letter after it had been written, but when a matter or the man addressed was of considerable importance he would *dictate* his letters to Thomas, who was in fact his personal secretary, in exactly the same way as a modern executive. In emergencies, or when addressing his equals, the duke would sometimes write a letter himself, and, as we shall see, he could do so with courtesy and finesse if the event called for it. The Gilberts, like the Brindleys, were originally yeoman farmers, but they had evolved into Staffordshire squires; they owned Cotton Hall, and their father had left a comfortable income of £300 a year. The boys were born at Ipstones in the same county,[23] and Thomas's legal training was complete by 1744, when he was called to the Bar.[24] A year later he was commissioned in the regiment mustered by Gower to resist Prince Charles and his Highlanders. Thomas Gilbert's immensely long parliamentary career from 1763 to 1795 as member for Newcastle under Lyme and for Lichfield was marked by a zealous and dedicated attention to the public interest.

The key to Thomas's character is inscribed on a stone outside his Chapel of St John the Baptist at Cotton in Staffordshire:

> Let future ages view this sacred place,
> And praise th'Almighty whose directing Grace,
> With Heavenly Zeal inspired the founder's mind
> To erect a Chapel to Reform Mankind.[25]

That founder was Thomas himself – sleepy Mr Gilbert, his fellow MPs unkindly dubbed him,[26] for an ailment of the eyes made his lids droop in an owlish fashion when in fact he was often quite wide awake. Though his aim of reforming mankind was never realised, he did have a measure of success in improving the Poor Law. A sincere and dedicated Christian, he was also a practical man endowed with an innately English distrust of theory and a dedication to the virtues of hard work and of good works. It was not surprising that Gower and

Bedford ensured that a man of this calibre should become the duke's personal assistant and frequent companion – none better for a wealthy and fatherless young man – but it is equally clear that Francis knew his own mind and made his own decisions from an early age. Thomas's parliamentary achievements included an Act for helping absentee clergy to live in their parishes by loans from Queen Anne's Bounty and valuable support for the newly evolving friendly societies which were shouldering a little of the social responsibility once undertaken in a more limited field by the guilds. He was much ahead of his time in championing the abolition of imprisonment for minor debtors and in 1787 obtained the Act for licensing dogs – an important advance in public health, especially in the prevention of rabies.[27]

Thomas was among the first to realise that, as the tempo of the industrial revolution quickened, so the parochial arrangements for helping poor, sick and destitute people were no longer adequate, and his 1782 Poor Law Act gave committees power to build workhouses which would be strictly administered. Though his achievement of encouraging two or more parishes to unite in aiding the poor (Gilbert's Act) was abused in Dickensian England, it was described as 'the foundation of many important items in Poor Law administration', while his publications on Poor Law reform were many and widely read.[28] All this was valuable work indeed, but his greatest achievement lay in serving as the duke's personal assistant and encouraging canal, mining and industrial enterprises, for the most practical way of reducing poverty lay in providing employment and prosperity where there had been none before. W. E. Tate wrote, 'the member for Lichfield had a genuine love of the poor and concern for their welfare'.[29]

John Gilbert, who was baptised at Alton Church near Cotton in 1724,[30] had none of the advantages of his elder brother. After some schooling near by at Farley, he was apprenticed at thirteen to the Boulton firm at Birmingham and grew up with Matthew Boulton of steam engine fame, who, though a little younger than John, remained his lifelong friend. In 1742, when he was only nineteen, his father died, and as Thomas could not abandon his legal studies in London, John looked after the Cotton estate, which included a lease of some valuable limeworks. By about 1753 he was already in the duke's service.[31]

By 1757, after completing legal work for the duke at Ashridge,[32] Thomas was free to make the family home at Cotton the centre of his activities, so in June John moved to Worsley to take up his main life's work as factor for that estate under his brother Thomas, the duke's General Agent. John's was a firm but diffident character, as befitted a steward, but his work was also his absorbing hobby. 'By profession he was a Collier Miner, Canal Navigator . . . a practical, persevering and industrious outdoor man, loved mines and underground works,'[33] wrote his chief accountant and personal assistant, Robert Lansdale, adding: 'In my youth I travelled with old Mr John Gilbert to all his various concerns in each county and assisted to examine vouchers, make up Books, Copy letters &c.'[34] He also had a gentle sense of humour and was engaged in such an extraordinary variety of

6 John Gilbert, 1724–95, resident engineer, and agent for the Worsley estate

undertakings at the same time that he was not only immensely busy but apt to be constantly in a hurry, though not, like his elder brother, quite so absent-minded about such details as documents.[35] John Gilbert had long been in the duke's employment, but it was during his survey of the Worsley mines about 1757 that the two men met most often, and the foundations of a lifelong friendship were formed. Gilbert's task lay in evaluating the factors governing the winning and carrying of this coal, and he talked to all who had any knowledge of the subject; he was lodging at the Bull Hotel at Salford, and, retiring to his room, he refused to see anyone for two days while he worked out his plans.[36]

As Gilbert pondered over maps and accounts it became clear that the most intractable problem of all was transport. The traditional outlet for Worsley coal lay south-westwards into Cheshire, around Lymm, Thelwall and Stalham,[37] and there it was that the Egertons sold most of their output, but it called for little mental agility to see that the really rich markets lay in the Salford and Manchester area, with their combined population of about 40,000. Moreover, work had begun on the Sankey Navigation on 5 September 1755,[38] which would have provided fierce competition around Thelwall for Worsley coal; it may therefore have been a shrinking market which drove the duke and his advisers to plan and execute his new canal. The Mersey and Irwell demanded tolls of 12*s* a ton for carrying from Manchester to Liverpool, and had failed to make the branch cut to the duke's mines. Even the recent improvements in the unsurfaced turnpike roads did not prevent them from becoming quagmires after heavy rain; the straining packhorses with their bobbed tails and tinkling bells could barely carry more than 280 lb each, while the country carts went creaking on their cumbersome way, seldom containing more than a ton or two. At Worsley pithead coal cost 10*d* a horseload, but by the time it reached Manchester shortages could drive people to pay almost double. It was just as bad elsewhere – around 1760 Josiah Wedgwood calculated that he was paying nearly as much for carriage as he did for coal.

Pondering these problems at the Bull Hotel in Salford, John Gilbert received that spark of inspiration which goes with so many new

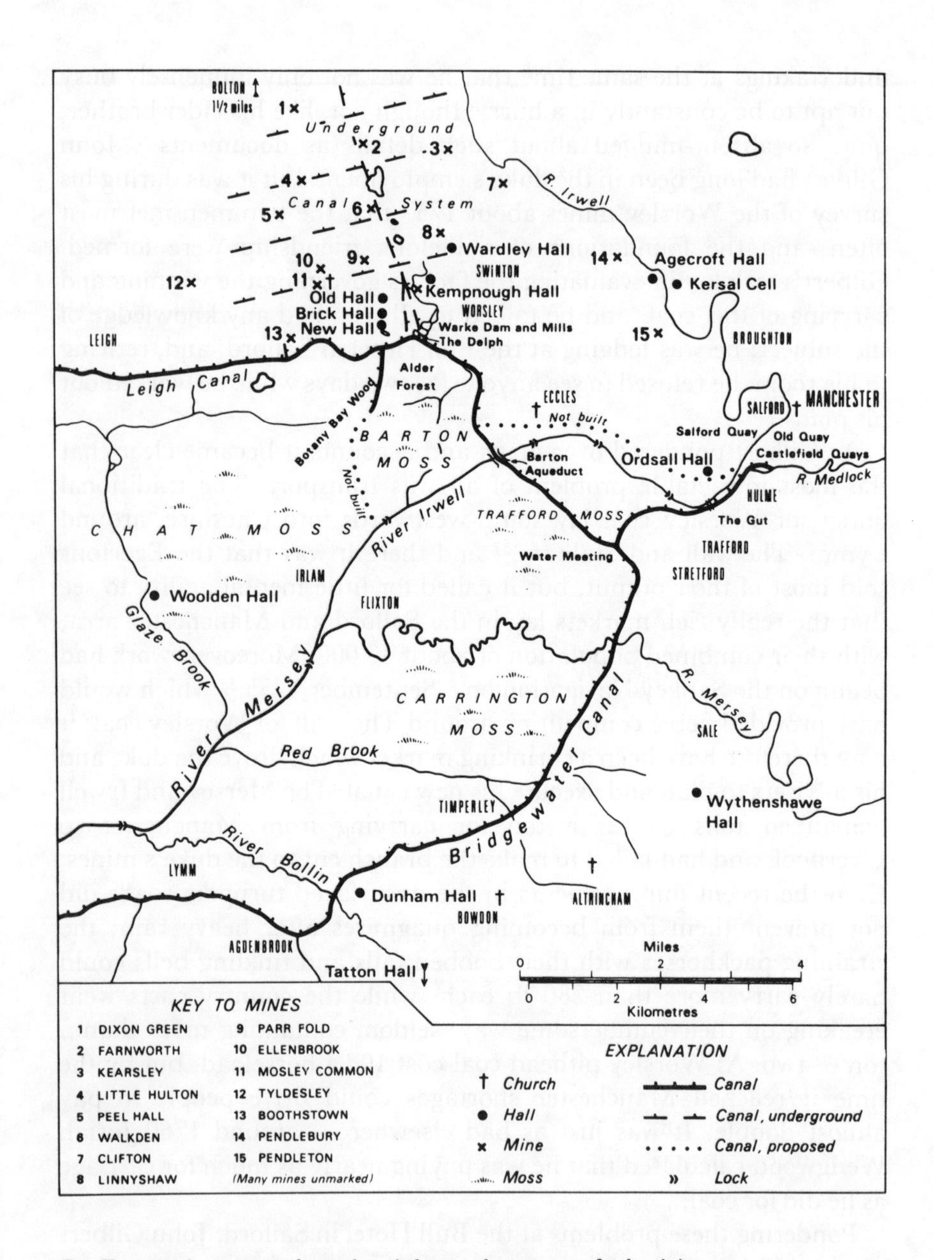

7 From mine to market: the eighteenth-century fuel crisis

inventions. No summit canal can work without an adequate water supply, and Scroop's draft agreement had indeed considered the use of sough water from the mines to float coal barges. Gilbert decided to carry this idea much further, and to him must go the full credit for solving three major engineering problems with one brilliant idea.[39] If a tunnel or sough large enough to carry *laden boats* could be drilled into the duke's mines from the Delph above Worsley Mills until it met his coal seams under Walkden Moor, these barges could carry coal from mine to market, the springs inside would provide a head of water for the canal, and the flooded coal mines would be really adequately drained by a deep rock sough, much less liable to collapse than the old one nearer the surface. All could be completed by one simple but very costly operation. 'The tunnel,' wrote Abraham Rees, a scholar of accuracy and repute, 'was entirely executed as well as planned by Mr Gilbert,' and again, 'as he preceded Mr Brindley in this business, of which we have ample and satisfactory evidence, we thought that justice required a candid and impartial statement of the case'.[40] 'Mr John Gilbert was in the employment of and domesticated with the late duke at the very commencement of making his Navigable Canal,' wrote the Rev. F. H. Egerton, and there is no evidence to the contrary.[41]

The influence of Samuel Egerton on his young cousin is more difficult to assess exactly, but it was certainly considerable, for it was he who was his guardian, took him in when he was homeless, taught him estate management, and indeed acted very largely in the place of his father. Yet one cannot avoid the conclusion that as the years passed the duke stayed more often at Trentham than at Tatton, while he preferred the house of a friend when the canal idea was beginning to take shape. Samuel Egerton was, perhaps, anxious to avoid a match between his young ward and his only daughter, Beatrix.

Last but not least, there was the duke himself. Suffering can be the beginning of wisdom, and Mrs Wood's letter shows clearly that he had matured into a man capable of making shrewd and effective judgements; never too proud to seek advice, the loss of Elizabeth Gunning had taught him something of humility, even though the lesson may not have lasted indefinitely. He had always been independent, introverted, capable of coping with his own interests, and possessed of a very powerful will, while his French affair had taught him discretion. Though he would never evolve into such a finely polished man of the world as Gower, with his magnetic personality, or Robert Wood, with his travelled and cultured insouciance, these were not necessarily the qualities required for canal construction and waterway politics. Certainly he was no boorish aristocrat, however much he might affect that part in later life, but a powerful and ruthless personality was evolving in conjunction with a fixed determination to serve the poor and provide employment. 'Of course,' wrote J. R. Harris, 'the Bridgewater scheme was in a sense above criticism – there was a recognition that the designs of the Canal Duke were inspired by something higher than profit. . . .'[42] The time had come for these designs, resources and characteristics to be deployed with a single-minded determination towards that end.

Notes to this chapter are on p. 183

5

The first Act

Francis was much impressed by John Gilbert's threefold plan for solving the Worsley water, mining and transport problem;[1] as he had inherited his extensive estates without legal reservations in 1757, the decision about excavating the barge tunnel lay with him alone. It was a momentous and costly one to have to take, but, as we have seen, he was ably supported by skilful agents and wise advisers. By October 1758 the duke was appealing personally to the citizens of Salford, and the larger town of Manchester also, pointing out that the 1754 project for a cross-country canal from Salford to Leigh and Wigan in Lancashire, via Worsley, had not been approved because it would have drawn summit water from the Irwell, robbing the old navigation and so creating a health hazard, while the undertakers had failed to fix a price limit on coal. Such a project was doomed to failure, but he claimed to possess 'a sufficient Quantity of coals to supply the consumption of Manchester and Salford', would build the 'Cut or Canal' at his own expense, supply it with water from Worsley Brook, and inform the public before applying to Parliament of the exact price at which his coal would be sold.[2] On 25 November 1758, therefore, 'A Petition of the Most Noble Francis Duke of Bridgewater' was presented to the House of Commons – waterway Bills almost invariably began in the Commons rather than in the upper House – so the canal idea was again launched, with more influential backing and a better chance of success than its predecessor of 1754, which had raised such powerful opposition from landowners and turnpike trustees.[3]

A petition was the customary method of opening an application for a private Act of Parliament, and in his the duke stated bluntly, and in the strongest terms, that the proprietors of the Mersey and Irwell Navigation had failed to meet their obligations to local coal owners: 'although upwards of twenty years are elapsed . . . no part of the said [Worsley] Brook had been made navigable, but the Undertakers have intirely neglected to carry the said Act . . . into Execution, whereby the intentions of the Legislature have been intirely frustrated and disappointed'.[4] The petition emphasised that the 'gentlemen, traders, manufacturers and inhabitants' of Lymm, Thelwall and other parts of Cheshire nearer Warrington, traditionally supplied by Worsley mines, had their coal intake crippled by dreadful roads. A survey clearly showed that a canal could be built from Salford to Worsley and on to Hollin Ferry, lower down the river navigation, for solving this coal supply problem, and for carrying general merchandise also. Francis demonstrated that he owned about a third of the land along the projected route, and was quite happy to build at his own expense as long as the canal became his property – a reasonable but controversial request. He argued that the canal idea would benefit more people than the Worsley Brook project could ever have done, while most of the landowners along the Salford route were happy to support him as long as they received adequate compensation. The petition was received by the House and sent for further consideration to a largely sympathetic committee of seven members, including Samuel Egerton and one of the Leveson-Gowers.

On 5 December 1758 the Commons received two new petitions vehemently supporting the duke, one from the gentry and tradesmen living in and around his main coal market at Thelwall and Hollin Ferry, and the other from Salford and Manchester folk. Both parties gratefully and genuinely acknowledged the vast reduction in fuel bills which this improved transport system offered, and supported it wholeheartedly, emphasising that the duke had promised to sell his coal at 'a Rate of Price not exceeding four pence per Hundred and that this would be of great advantage'.[5] The House still seemed more than a little dubious of the extent to which private gain balanced the public interest in this new project; John Locke's philosophy, with its emphasis on the inalienable rights of private property, was increasingly influential, and many Members appreciated that the sanctity of private property here lay much at risk. It was one thing to pass Acts for navigations which followed the beds of rivers, or for turnpike roads which sheep and cattle could cross safely enough in those days, but quite another to delve canals in other men's land dangerous to lives and livestock alike. This was no era of common rights, but an age of walled parks, stern gamekeepers and judicial transportation. From the duke's first Act evolved the right to compulsory purchase of vast areas for railways and motorways – a quantitative innovation of considerable importance in British history, though not a legal precedent, since navigations had done this earlier.[6] But the squires who rode to Parliament through so many puddled miles were well aware of the need for better transport – they remained sympathetic, and the Committee was enlarged by two City aldermen and all the Members for Lancashire.

By 6 December 1758 the House agreed that the canal idea was worthy of further investigation[7] and William Tomkinson, the duke's Manchester agent and solicitor, opened the case for his client. Tomkinson emphasised that he had paid a visit to Worsley Brook only a fortnight ago, and could vouch that nothing whatever had been done to make it navigable in accordance with the terms of the 1737 Act – a telling point, this, because it implied that parliamentary time and trouble had been wasted on an Act which the Old Navigators, as the supporters of river transport were nicknamed, had completely failed to execute. He made it plain that the transportation and price of coal were the root of the problem, for it cost a minimum of $5\frac{1}{2}d$ a cwt at Salford and Manchester. Water supply was, he admitted, a problem, because the Navigators clearly would not wish to lose theirs, but he emphasised that Worsley Brook was fed from the Old Sough which drained the duke's own mines. It was a factual pleading, but telling in all its points.[8]

He then handed over to the Resident Engineer, John Gilbert, who produced his map of the Salford–Worsley–Hollin Ferry canal, stating that he had 'attended at the Levelling and measuring of the ground'.

Levelling is a term still used in surveying, and simply means establishing a just level for constructing any engineering or building project; it does not have to be, and in tramroad and railway work seldom was, horizontal, but rather at a slight angle to allow for inclines.[9] Gilbert was firmly convinced that a canal was a feasible and practical proposition capable of carrying 'Vessels of considerable Burthens, whereby the carriage of goods will be greatly facilitated, and become less expensive'.[10] He too emphasised that there was no question of or need for stealing the Old Navigators' water supply, because nearby Chat Moss, a swampy morass, Worsley Brook and other springs on the route could provide a plentiful supply. He added that the duke would be willing to bind himself to sell his coal at the remarkable saving of 4*d* a cwt, while the new canal was, he pleaded, better, more beneficial to the public and preferable in every way to the old idea of making Worsley Brook navigable by building vastly expensive locks down a fall of 40 ft or more,[11] a project which coal alone could never make viable, while boats would work far more quickly on a lockless canal.[12]

Gilbert then quoted statistics assembled by his staff to prove that, while the duke owned about a third of the land on the projected route, rather more than another third belonged to people happy to grant permission for a cut through their land, while only a very few – the proprietors of 3,200 out of the overall 26,991 yards – had actually refused their consent. Such evidence was, again, factual and impressive enough for the House to agree to bring in a Bill which was sent to a far larger committee of sixty, and all the Members for 'Lancaster, Chester, York, Derby and Stafford'.[13]

The Old Navigators did not openly oppose the Salford Canal Bill, partly because they had failed to make the brook navigable, but chiefly because they stood to benefit from the duke's plan to link his waterway with their own by 'making a Communication between the said Cut or Canal and the River Irwell betwixt the Lock called Holmes Bridge and the Sugar House situate on the banks thereof in Manchester'.[14] This was close to Ordsall Hall in Salford. When this agreement was eventually signed by Tomkinson and the Old Navigators on 14 June 1759 it gave the duke the right to trade on the

Irwell at the preferential rate of 6*d* a ton,[15] on condition that his vessels gave way at all locks to loaded boats and to barges paying a higher tonnage.

The sanctity of private property still remained the weakest link in the duke's Bill, and there the opposition concentrated their counter-attack. On 5 February 1759 a petition from owners of land between Worsley and Salford advanced a strong case, arguing that the new canal would be 'very prejudicial to the respective lands and properties of the Petitioners'.[16] The opponents included Christopher Byrom, and Mrs Chetham, a wealthy woman, whose solicitor wrote early in 1759 that the landowners who were willing to join his client in opposing the cut were 'now so few and so terrified with the prospect of Expence and hazard of Success' that he doubted if she would continue the struggle, adding that the Old Navigators and others were unwilling to lend their support.[17] Though depleted, the opposition gathered sufficient strength to plead a case, their counsel doing his utmost, but on 23 February 1759 Mr Rigby, another friend of the duke, reported back to the House that, while the committee had made several amendments, it was clear that it supported the canal idea in principle. The public interest had proved stronger than the arguments for private property: a tolerant and far-sighted decision for a parliament of squires.

So the Bill passed to the Lords, where the young duke was waiting for it impatiently.[18] Lord Hay stated that George II had granted his approval in principle as far as the Crown's interests were concerned – Francis had probably explained his motives, and the virtues of the new canal idea personally to the King. A committee was formed to assess the virtues of cross-country waterways, with the duke himself a member,[19] aided by the Lord Chamberlain, fifteen earls, including Stamford, who was a neighbouring landowner in Lancashire, three bishops and six peers; Viscount Weymouth was in the chair. John Gilbert was called to give evidence before this formidable array and did so in an understandably shy and temperamentally laconic fashion:

He is asked what his Employment is.
Says he has many Employments, but his chief one is being concern'd in Lime Works.

He is asked what ye cut will be of [*sic*]
Says, for carrying Coals, Goods &c. for ye use of the Country.
He is directed to withdraw, and His Grace ye Duke of Bridgewater is present & consents to ye Bill & prays ye passing thereof.[20]

On 12 March the question was put to the Lords of whether the Salford Canal Bill should pass, and it was resolved in the affirmative; the duke was present again on 23 March 1759 when the royal assent turned the Bill into an Act with the ancient Norman declamation *le Roy le veult.* Clearly Francis had prepared much of the ground himself, particularly in the Lords, and though the opposition had not been powerful enough to make it a very costly Act – perhaps a few hundreds from the duke's pocket – this was because they could win only limited support against a plan so obviously beneficial to the public, and from which even they might gain. More costly, vicious and hard-fought battles lay ahead, but without ducal patronage and wealth behind it the Act could well have suffered the fate of the abortive 1754 canal. Considering that private property was at stake, it had been passed very quickly by any standards, demonstrating the considerable power wielded by the Bedford–Gower–Egerton interests in both houses. The sympathy felt for people like Mrs Chetham whose land would be cut through could scarcely outweigh the benefits from vastly reduced fuel prices, combined with the value of carrying general merchandise and agricultural produce by swifter and more reliable routes and methods.

The Duchess of Hamilton married Colonel Campbell, future Duke of Argyll, in March 1759, and on the fourth of that month, just before his Bill became an Act, Francis gave a magnificent ball in London to which the whole world of fashion was invited. While he owed an accumulation of hospitality from the busy years of canal preparation, he was doubtless in a mood to celebrate, as there was little chance of opposition in the Lords after the third reading.[21] Walpole's sister, Mary, Lady Cholmondeley, was there, discussing her host behind her fan with a good deal of sympathy, but Francis knew that even if the gamble he was embarking on did eventually prove a success it would be many years before he could afford to support a fashionable wife. So convinced was he that his Bill would pass that he had ordered 219 printed and stamped blank deeds of lease and release to be sent to Warrington early in 1758,[22] but all his achievements were enacted against a sombre background of personal tragedy. His sister, Diana, Lady Baltimore, had died of the family ailment in August, leaving only Francis, Anne, Caroline and Louisa out of Scroop's eleven children, and, of these, two more would be dead within five years.

On 22 October 1758 the duke had written personally to landowners along the route of the Salford Canal, sending them printed leaflets:

> the inclosed scheme having met with encouragement from Gentlemen and Traders in Manchester and Salford, I take this method of communicating it to you as a proprietor of lands thro' which the Cut is intended to be made, and if it shall receive your approbation and Concurrence it will much oblige your most obedient Hon*ble* Servant Bridgewater[23]

Even when his Act had been passed the insecurity of the scheme was emphasised by a clause insisting that, if the duke was unable to pay full compensation for the lands he had expropriated, former owners could levy on his coal sales sufficient sums for recovery. The clause limiting his sales in Salford to 4*d* a cwt for forty years, though doubtless necessary to get the Act through, was crippling, while another limited his toll to a ceiling of 2*s* 6*d* a ton, though he later benefited from a far-sighted permission to charge for goods left more than twenty-four hours at his warehouses. Historians have harboured such a determined antipathy to the duke that it is impossible to find any acknowledgement of his generosity in fixing his prices lower for coal at $3\frac{1}{2}$*d* a cwt for many years, until inflation and increasing debts drove him to exceed his limit – an action for which he has been more justly criticised, since he should have made his peace with the law in this matter.

The duke's motives

It is more difficult to assess the extent to which the duke and his advisers were motivated by benevolent and idealistic intentions. His introduction had claimed that the canal would be 'Beneficial to

Trade, advantageous to the Poor . . . and answer many other good purposes', and later in his life he made it a conditon of his support for the Trent and Mersey scheme that it too must be beneficial to the public, at a time when he was not yet certain that his own waterway would prosper, for he was what was later termed a Christian Utilitarian in philosophy, if not invariably in his actions. It would have been a hard man indeed who could have remained unmoved by the horrors of any eighteenth-century town, and the Manchester Constables' Accounts show that area to have been no exception to a rule of prostituition and poverty, forgery, superstition, vice, gin-alley drunkenness, public floggings and hangings and other Hogarthian horrors, however much balanced by corresponding virtues of faith, hope and charity. One critical factor was simply to get people jobs – to provide employment – and Francis had suffered enough, and recently enough, in his childhood to feel some sympathy for the underdog.[24] In the light of new documentary evidence it is clear that the recent denigrators of the duke's character were very generally, if not invariably, wrong, while Professor W. T. Jackman's penetrating analysis was, after all, probably correct when he wrote that Francis 'turned his attention to the relief of the social and industrial life of the city of Manchester and its surrounding towns'.[25]

Crippled by such stringent price regulations, a Salford Canal running alongside the Old Navigation was hardly a paying proposition in the foreseeable future, so Francis, Gower, Samuel Egerton and the Gilberts began 'considering the problems of communication between Worsley and Manchester in the first half of 1758 at the latest'.[26] While it would be a mistake to think of the duke's first Salford Canal Act as a vast and costly deception, it is now clear that he and the Gilberts are unlikely to have made a final decision about the route until work had started and the many technical difficulties of labour, engineering and water supply had begun to be ironed out. It was not until 1761 that Francis's survey of his twelve estates made it absolutely clear how much income was available for his canal projects.[27] Earl Gower, a man capable of long-term strategies, had paid Brindley to spend a fortnight riding the Trent and Mersey line in February 1758, and Thomas Gilbert was his agent as well as the duke's, while Samuel Egerton would surely have been consulted. There is now definite evidence[28] that Francis was considering the idea of taking over the carrying trade between Manchester and Liverpool, potentially a vastly rich prize, from the Old Navigators, and linking with the Trent and Mersey, as early as 1758, though he probably did not make the irrevocable decision until the cut drew near to Barton, beyond which any work on the Salford route would have been so much labour wasted. The idea of a canal network reaching across England and linking key ports with water transport goes back at least as far as Francis Mathew's attempt to interest Cromwell in a London-to-Bristol route via the Isis and the Avon,[29] and was common knowledge among the squires and merchants who passed the innumerable navigation Acts of the eighteenth-century. There can be little doubt that the Gowers and Egertons were among those who were considering a national canal system long before Brindley was engaged on the Worsley project, and linking the Trent to it.

By March 1759, and probably well before that, work had begun,[30] with Gilbert's men hammering away at the underground canal from a quarry called the Delph just upsteam from Worsley Mills. As in the case of the Dutch city of that name, the word derives from *delve*, and is still used locally for entrances to shafts, quarries and soughs. Mining tools were primitive in those days, and the wooden picks with steel heads needing constant attention made scarcely any impression on the rock under Walkden Moor; the miners grew weary of the seemingly pointless task and abandoned work. Francis, it was said, managed to convince them by taking out his snuff box, holding up a pinch, and promising that he would continue to pay wages for as long as they could extract even that much from the sough. Since gunpowder was already used extensively in mining, this tale may have been exaggerated.

Notes to this chapter are on p. 185

6

The canal triumvirate

Francis was only twenty-two years old when the canal work began – an impatient young man, something of a hustler, always anxious to speed things along. Had his aims been less benevolent he might have gained a quicker return on his capital outlay in slaving or sugar, the spice trade, cotton or government contracting, but his training was in estate management and he came from a line of responsible landowners anxious to get coal away to good markets and provide adequate employment. He came also from a practical breed which could not long tolerate an outlay earning no return at all.

The older inhabitants of Worsley must have been shaken by the growing army of men pouring into the new works. Skilled boatbuilders like James Lomax were plying their craft by August 1759 for 21*d* a day, while companies of navvies like Daniel Kay and his men were paid piece-work for delving six roods at 10*d* a day each. Absurdly low as such wages may seem to us, they were considered fair by contemporary standards. John Gilbert moved house to Worsley on Midsummer's Day, 1759,[1] but he had been in charge of operations there since about 1753; his work included bargaining with local squires and farmers for land taken for the cut, opening new account books for recording the wages of blacksmiths, masons, bricklayers, miners, and many other craftsmen, settling the men's working and family problems, accommodation and finances, and seeing that the line of the sough ran true from the Delph beneath the hill, while overseeing even the minutest details with patient care.

The year 1759 was auspicious for Britain. In August the threat of invasion was dispelled by Boscawen at Lagos, and in November Hawke pursued a French fleet among the reefs of Quiberon Bay to win a resounding victory in a full gale on a lee shore. In Canada Wolfe stormed the heights of Abraham, and in India Clive's earlier triumph at Plassey had been followed by the capture of Masulipatam, which substituted British for French influence at the court of the Nizam of Hyderabad.

At home there were various changes which, though they seemed insignificant or scarcely perceptible at the time, were soon to transform agricultural England into industrial Britain. In 1759 Josiah Wedgwood, twenty-nine years old, began business on his own account in a humble cottage near the market place in Burslem. In the same year Matthew Boulton's father died and within three years he had built the Soho factory at Birmingham, where later the famous steam engines were assembled in partnership with James Watt, to drain the mines and, later still, drive the mills.[2] For half a century the English iron trade had been developing by substituting coke for charcoal in the blast furnaces of Coalbrookdale. By about 1767 Arkwright was inventing a frame for performing all the operations of spinning at once; gradually this began to supersede the cottage spinning trade based on out-work and suck manufacturing into the smoking cities.

King Cotton had not yet grown to rule unchallenged in Lancashire,[3] for hand-loom weaving of woollen cloth was still a popular cottage industry around Worsley in Scroop's time, while there was also much weaving of mixed cloth like the strong fustians with their linen warp and cotton weft, and though some of this was sold locally a good deal had long been exported. Liverpool was the nearest port, but the dangers and vagaries of sailing ships combined with the tricky entrance to that harbour[4] meant that some goods were despatched to Bewdley on the Severn, Winsford on the Weaver, or Wilden Ferry, now Cavendish Bridge, near Shardlow on the Trent, for forwarding in barges. The Egertons and Gowers were aware of this absurdity by which Manchester goods were being shipped through Bristol, still the most important port on the west coast. Manchester had been 'long noted for various branches of the linen, silk and cotton manufactory',[5] but cotton was well in the ascendant as the easiest, cheapest and most popular fibre, soon to be carried so extensively on the duke's cut.

Brindley's arrival

About May 1759 John Gilbert, ever on the lookout for talented engineers and workmen, introduced James Brindley to the duke and persuaded his patron to engage him for a while.[6] Schemer Brindley, as his friends had nicknamed him for his inventiveness as an engineer, had already, very briefly, reconnoitred the Trent and Mersey for the family[7] and had then been engaged in constructing the 'Dam or Cascade at the foot of Trentham Lake', and it may have been this particular work, in addition to his growing reputation, which encouraged the other two members of the triumvirate to engage him for canal construction.[8] Moreover, Gilbert was overloaded with work and was beginning to delegate authority; as the duke was often away dealing with canal politics or his other estates, and sometimes racing at Newmarket, Gilbert had a wide discretion in minor matters, though questions of policy were dictated by his employer to Thomas Gilbert in letters of which only a few have survived.[9]

With the arrival of Brindley at midsummer 1759 the canal story takes a new and dramatic turn, for here was a man who was not only a 'character', but also, in his forcefulness, ingenuity and intuition, worthy of such a noble enterprise. Few men indeed have led such a constructive life as James Brindley, but he has been honoured for what now seem to have been partly the wrong reasons. True, G. M. Trevelyan described how Francis 'allied his Parliamentary influence and his capital to the genius of his half-illiterate engineer', while H. A. L. Fisher wrote of how 'William [*sic*] Brindley, an illiterate genius, engineered the Bridgewater Canal', but these statements, deeply though they have penetrated the general and academic curricula, were based on a series of errors by Smiles, Phillips and other writers who were not able to consult the documentary evidence. Brindley's great achievement lay in carrying out the canal idea around England, but there is no evidence to suggest that he was the originator of the Bridgewater Canal either above or below ground, or that he was ever in command of policy, however much he may, as the main consulting engineer, have contributed towards it. What we can claim with reasonable certainty is that Brindley did survey some of the Bridgewater Canal, though not the earlier section, under Gilbert's supervision,[10] and that he probably worked fairly consistently on it for about three of the seventeen years that it took to construct it,[11] He may have had some influence on policy when he was the duke's guest at Worsley and at the cottage which the triumvirate used at Barton, but there are indications that he caused almost as many differences as agreements of opinion.[12] Though he invented a most ingenious sand-shifting gearing for Worsley Mill, may have helped to build water-driven bellows for pumping air into the mine, and improved the cranes at the Delph, his system of rising gates for blocking off the canal when banks burst worked better in theory than in practice, though it did have some propaganda value in reassuring the unfortunate householders who lived near by.

The Brindleys, Brinleys, de Brundeleys spelt their name in different ways, but came from the same clan which spread out into Tasmania and Philadelphia and has indirect descendants in England.[13] Brindley's father came from a line of independent, fairly well-to-do and sometimes literate yeomen farmers, though he may have been a rather unstable person, and James's mother had to teach her son to read and write – not an unusual arrangement in those days. The boy loved mills and water from the earliest age, and in 1733 began a typical apprenticeship under Abraham Bennett near Macclesfield in Cheshire, but his ability and honest, enduring workmanship enabled him to set up his own business at Leek in Staffordshire.[14]

In 1752 Brindley began an engineering project of quite exceptional ingenuity. Wet Earth Colliery, a mine between Salford and Bolton, was drowned out, but he managed to use the latent power of water to conquer water, by building a very long mill race. The lie of the land obliged him to take it nearly 800 yards underground until, by digging a syphon, he carried it right under the river Irwell and up the other side. Thence the race ran on until this considerable head of water met a breast wheel, driving pumps to drain the mine dry. Parts of this brilliant engineering achievement had a life of 175 years.[15] There were further outstanding technical inventions at a Congleton mill which were later adapted effectively to the cotton industry and these, combined with his experiments with a 'fire injon', as he graphically described what we now call a steam engine, do entitle Brindley to a niche among the pioneering engineers, though his greatest achievement lay in spreading the canal idea. He was to some extent handicapped by a limited education and outlook, and if he lacked the temperance and wisdom of his friend Josiah Wedgwood, or the polished efficiency of Matthew Boulton, when it came to handling water and surveying, or inventing new mill machinery, he seems to have had no peer.

Brindley's limitations

A miniature owned by Brindley's descendants shows him wearing a brown tye wig, a well cut snuff-coloured coat and a white waistcoat and cravat. All very smart, but underneath lay a tough master craftsman turned engineer, an attractive and domineering personality – tetchy but imaginative. His inventions could work spectacularly or fail dramatically, evincing in his diary such self-depreciatory jottings as 'April 21 1758 to run about a Drinking 0=1=6',[16] or that later icy entry in the Oxford Canal's Minute Book – 'the Engineer, Surveyor and Clerks of this Company do not associate or drink with any of the Inferior Officers or Workmen'.[17] Like so many in Rowlandson's age he was a gourmand, and ate heartily until his expanding stomach hoisted a button on his waistcoat to a specific point beside his coat, when he would immediately call a halt with engineering precision. There was something Falstaffian in this irascible engineer who, given a drink or two, was not only humorous himself, but 'the cause that humour is in other men'. He lived more in the oral than in the written tradition, and could be a brilliant advocate of any project like canals which won his warm-hearted conviction; he also treasured a little library of books.[18] His one surviving map demonstrates that he could never have drawn up mining plans, or stress diagrams for Barton aqueduct,[19] but a touch of poetry has left us with a few sayings, chief among them that water is like a giant, safe only when laid upon its back. He had a natural son called John Bennett – a union far from unpropitious, since the novelist Arnold Bennett was directly descended from the old engineer, and inherited his narrative gifts.[20]

When Brindley rode up the hill to Worsley on 1 July 1759 it is just possible that he might have visited the work earlier,[21] but there is no evidence. He came as a consulting engineer-millwright, at the head of his own small company of craftsmen, recorded in Gilbert's accounts for their removals: 'Pd Messrs Brindley, Geo Harrison, Saml. Adam and Saml. Bennet's expenses with their family £1–9–6'.[22] Brindley himself was treated as an honoured guest, with free board, stabling and expenses, proving that his views on surveying and water control were of some moment to the two younger and, in water supply, less experienced men. Francis later gave him an inscribed and signed prayer book, suggesting that conversation at the Old Hall sometimes turned to subjects other than canals.

This new triumvirate formed an able team for the formidable task of providing employment in the Manchester area. Francis was already a competent and well trained estate manager, wielding considerable political influence in conjunction with Gower, and backed by broad acres of landed wealth. John Gilbert was a trained engineer of genius and a brilliant land agent, administrator and accountant, while Brindley had rare gifts when it came to machinery and water, or advocating canals in or out of Parliament. The duke also saw to it that he was very well paid, contrary to the myths of some historians.[23] As Resident Engineer and agent for the estate Gilbert had to work very hard, though aided by several able assistants, and received £200 a year, which soon rose to £300.[24] The duke, with his many unavoidable outgoings, had cut his expenditure back to £400 a year, selling his hunters at Ashridge and employing only a valet and a groom for his personal needs. Under such circumstances a consulting engineer who sometimes worked away at his own practice could not complain of an initial £119 which was rapidly increased, and despite the protests of his widow, he never did so. It was not many years since Samuel Egerton's brother, Thomas, had scraped by on £40 a year as a junior clerk in Holland.

As the triumvirate laboured on, the cut began to wind away from Worsley in three different directions: down towards the Irwell at Barton, along the line of the underground tunnel or sough, and westwards to Chat Moss, where waste could be dumped to firm up the morass. But then came a dramatic change of plans. Francis and his advisers had made up their minds not to run the canal to Salford parallel with the competing Mersey and Irwell after all, but to build a barge aqueduct at Barton, carry the cut high across the old navigation, and make a bid for both the coal market of Cheshire and Manchester, and for the Liverpool carrying trade.

The second Act

Brindley's gifts as an orator had probably more than registered with his colleagues, and on 23 January 1760 he noted in his diary that he rode off for London. On the 25th the duke presented his petition for a new Act, stating that he had already built part of his canal and in order to 'drain his Coals and establish a lasting Colliery for the purpose of supplying Manchester and Salford with Coals . . . hath begun a Sough or Level . . . and is now carrying on the same at a great Expence in a very Effectual Manner'.[25] But he added that it would 'be more convenient to the publick and less difficult and Expensive to your Petitioner to have the said . . . Canal carried over the River Irwell at . . . Barton Bridge', and so to Manchester, with a branch line for supplying Stretford and 'a considerable part of the County of Chester.'[26] Quite how much he planned to invade was not apparently guessed as yet.

This was the first barge aqueduct in England (and there is no evidence of one in Roman times) planned and submitted to Parliament for approval. After initial consideration by a committee, of which Alderman Dickinson was chairman, on 31 January 1760 James Brindley came forward to be examined. He stated that nearly three miles of canal had been made already and 150 yards of sough measuring 7½ ft high by 5½ ft wide.[27] He claimed that the Salford Canal presented grave technical difficulties – particularly rock cutting for over two miles[28] – arguing that the new route offered better water supplies. He added that there would be 'no difficulty in carrying the Canal over the River Irwell, near Barton Bridge', a prophecy he lived to rue; nor would the navigation of the river be impeded if boats were haled up-river on planks under the arches. It seems likely that Brindley, with his round, innocent-looking eyes and broad Derbyshire accent, was making something of a case here, for the technical difficulties of the Salford route were not insuperable,[29] but the new markets in Manchester and Cheshire offered the triumvirate a better prospect whether or not the duke's funds could run to an extension which would capture the Manchester–Liverpool carrying trade and link with the projected Trent and Mersey

After a second reading the Bill went to a committee of forty-one Members and all MPs for Lancashire, Cheshire, Staffordshire and

8 Barton aqueduct, painted by J. J. Rousseau

Yorkshire, following much the same pattern as the duke's first Act, but then it ran into the stiff opposition of a counter-petition from the Stretford to Manchester Toll Road Commissioners. They had just borrowed some £3,400 on the credit of forthcoming tolls to improve their muddy highway and raise substantial embankments against the Irwell's treacherous floods. Greatly alarmed by the prospect of canal competition, they argued that falling tolls would kill all hope of keeping their road in decent repair, maintaining that swivel bridges on the cut could only delay the postal service.

Though these were powerful arguments, they were countered by a forceful and well timed petition from the united gentry, tradesmen and inhabitants of Altrincham, Nether Knutsford and Stretford, in Cheshire, strongly supporting the duke's canal, which could supply their coal at such a vastly reduced cost. As the struggle wavered Brindley was called to plead before the committees. Finding the technical details somewhat incomprehensible to various Members, he carted a large cheese into Parliament and carved it into a very tolerable working model of a barge aqueduct. With this ocular demonstration, as he termed what we should now call a visual aid, he was able to give them a general idea; but some, it was said, were still unconvinced that a large stone trough could be made to retain water. At the next session Brindley carted in sand and clay and a large can of water. Forming the broken clay into a rough trough under the startled eyes of the legislators, he emptied half the water over it, and naturally enough it flowed on to the floor. Then he kneaded the sand and clay with a little more water until it was properly puddled, and again shaped out his trough. Refilling his can with water and the room with the patter of a professional conjurer, he poured a liberal libation into his model, and behold, instead of spilling, the water held, and he was able to explain that that was how his Grace the Duke of Bridgewater intended to convey his canal high above the old navigation. Fortunately the comments of the charladies who cleared up the mess the following morning remain unrecorded in parliamentary archives.[30]

With advocacy such as this it is not surprising that the Commons should have approved the Bill, which was carried to the Lords by

Alderman Dickinson, and once there the duke feared no further opposition from his peers. There were amendments in the Commons, but none by the small committee in the upper House, which had recorded the royal assent by 12 March 1760.[31] On balance it would seem that this second Act was mainly a *variation* Act, and that Gilbert and the duke did not suffer undue anxiety about the risk of failure. The Old Navigators may have regretted the loss of the Worsley coal trade, though they were probably thankful to be spared competition in other merchandise, but the alarm of the Turnpike Commissioners was only partially justified, and unlikely to impede a work so clearly beneficial to poor and rich alike. Though Brindley appeared as the chief advocate for this particular Act, the bulk of the administration had falled on Gilbert, who was paid £150 for his extra work on these first two parliamentary Bills.[32]

Meanwhile the cut was running into difficulties as it advanced towards the Irwell. On 10 November an embankment burst, sweeping a wheelbarrow so far away that the man who rescued it was paid half a day's wages for his trouble. Accidents of this kind meant that land gouged or flooded had to be made good and proprietors compensated for damage.[33] Though the duke, Gilbert or Tomkinson negotiated the purchase of land between Worsley and Manchester before work began on the channel, difficulties sometimes developed, and a body of canal commissioners was formed with power to empanel a jury if the need arose.[34] The duke was obliged to be fairly generous in his compensation, more generous perhaps than the modern official expropriators, for if the sum agreed by the jury was higher than that settled by the commissioners the duke was obliged to pay the jury's expenses – a very concrete obstacle to costly and time-consuming litigation[35] – so naturally he often found it cheaper to pay extortionate prices. The general principle was that, while the law would not normally force a man to sell his garden, it could oblige him to surrender agricultural land.

Just as in later years the railways used the canals to feed material for construction to the railhead, so the duke used the old navigation to bring up timber, stone, bricks, lime and other heavy goods to Barton, where, by June 1760,[36] the long embankment was rising and the three sturdy arches were beginning to grow out of the bed of the Irwell. The canal was sealed at the aqueduct workings, and a gig crane installed on the parapet to lower the twig baskets on to the duke's river barges, which had already brought up the heavy goods, so that Salford and riverain Manchester could be fed immediately with Worsley coal.[37] By July the duke must have been pleased with progress, as he ordered 'Ale to the workmen at the Bridge a quart apiece',[38] followed by more free issues of drink to the boatbuilders, Richard Mugg and John Lloyd, at a launching ceremony. The earlier agreement for a communication with the Old Navigation stood, but the site was changed to Cornbrook, near Manchester, where a short canal known as the Gut went down to the Irwell, so that boats built at the Worsley yard could pass either way, though this could not operate until 1763 when the main line arrived there.[39]

Brindley was in charge of Barton aqueduct under the general surveillance of John Gilbert, who doubtless drew the engineering plans, since his colleague and consulting engineer was never a competent draughtsman, though he was popular with the men, knowing exactly how to allay their many superstitions and sympathise with their problems and prejudices. Accidents, burst banks and sickness were bad enough, but to these must be added the frustrations of working with, for example, the foreman bricklayer, who invariably 'ruled the planets to find out the lucky days on which to commence work'.[40]

Notes to this chapter are on p. 186

7

The third Act

Fine fortunes are often won by brave borrowing, but the problem facing the duke was to persuade people to lend to an enterprise judged to be more an act of *noblesse oblige* than practical business in the Manchester manner, while waterways like the Mersey and Irwell, which had paid no dividend for the first twenty-eight years, made the major investors more than hesitant. Rather than form a joint stock company, Francis was determined to treat the canal as an extension of his Worsley estate, so he sent his agents riding far afield to beg money from individuals in return for bonds. These were stamped legal documents stating the amount of each loan, the name of the investor, and a rate of interest ranging around four per cent – they were generally redeemable at will.[1] The security he offered was his own titles, and twelve broad estates, but the terms of the Worsley entail excluded any mortgaging there, while some of his other properties were in various ways encumbered. Poverty is a somewhat comparative affair, but Lord Ellesmere's later statement that Edward Leigh, a yeoman farmer at Worsley, was typical of the poorer people who gladly lent their small savings for the canal, proves that capital was desperately hard to come by.[2]

By 1762 the duke had already spent some £28,000, as Gilbert records in his fascinating *General State . . .*,[3] the central account book in Worsley estate's vast accounting system, which, it must be remembered, was only one of the twelve main Bridgewater properties. This ledger provides, from 1759, a panorama of the canal, colliery, lime and farm concerns, though it does not allow for the loss of capital by coal extraction, and it shows the gradual building up of an immense debt until 1786, when the tide at last turned as these undertakings began to show a small but gradually increasing profit. It does not detail the income which enabled the work to be financed.[4]

One day, as John Gilbert was riding alone to beg a few more pounds from some of the remoter tenants, he was greeted by a plausible stranger on a magnificent horse which he was very anxious to exchange for the one which Gilbert was riding. Clearly the other mount was in far better condition, and so the bargain was agreed. Hungry and thirsty, the agent drew rein after a while at an inn, where the landlord greeted him profusely and asked him if his saddle-bags were well filled. Gilbert was puzzled at first, because he had never met the man before, but it gradually dawned on him that he had changed horses with a famous highwayman whose steed had become so well known that he was anxious to be rid of it.

Francis had employed Robert Fowler (1726–1801), later Bishop of Killaloe, Archbishop of Dublin, and an Irish Privy Counsellor, as his private chaplain until 1756, by which time he was beginning to make savings for the canal. Fowler, who then became a chaplain to George II, was noted for a kindness and affability not 'unattended by warmth of temper',[5] combined with a very solemn manner of reading services which probably amused his youthful patron, but it may well be that this distinguished churchman had a valuable influence on the duke's determination to provide employment, at a time when he was sensitive to advice.[6] While there were several inspiring exceptions, the Anglican clergy of Wesley's age, though often sound scholars, were not noted for consideration for the poor, and the duke, whether justifiably or not, felt that he himself was shouldering the heavier burden. He directed so much of his tithe money to the good work of canals that he caused much aggravation to the Rector of Eccles, who was naturally as anxious to pin his patron down as Francis was to evade him. One day the rector lay in ambush near Worsley, but Francis turned and bolted, so there followed one of the strangest hunts in Church history, a priest in pursuit of his ducal patron. The lungs of righteousness proved the stronger, and though Francis flung himself, panting, into a sawpit, he was run to earth. When he managed to get his breath back he promised full payment, and did issue the money soon afterwards.

Barton aqueduct

Meanwhile the aqueduct at Barton was nearing completion, while a labour force of about 400 men delved the channel across Trafford Moss, a quagmire so deep in places that cattle had been lost in its oozy depths. At last Brindley claimed that the aqueduct and southern embankment were ready to take the weight of water, and as a test the great trough was flooded. To everyone's horror one of the arches showed every sign of buckling. Poor Brindley, who had been drinking rather heavily for some time and overworking for weeks, had a nervous breakdown and retired to bed at the Bishop Blaize tavern at Stretford, while Gilbert came to the rescue. Brindley had laid too much weight on the sides of the arch, but Gilbert removed the clay and puddled it again.[7] When this was safely done the duke ordered a gargantuan feast for the men who had laboured under such miserable conditions through particularly wet weather, while they claimed that they had 'crowned the arches', since it coincided with George III's coronation.[8]

It had been a close shave, but after Gilbert's ministrations the masonry held firm, and on 17 July 1761 Francis invited his friend, Lord Stamford, and other guests to see the water in the first English aqueduct rise gradually to the level, allowing a flat carrying some fifty tons to be towed across. To the modern mind the triumvirate's achievement seems in no way startling, but to an untravelled contemporary visitor it had much the same impact as the space programme – 'perhaps the greatest artificial curiosity in the world', commented one observer, with insular ignorance of earlier works on the Continent. Crowds flocked from all over the country, and it became fashionable to visit Barton and the sough which was being drilled at Worsley. By any standards the duke's work was an impressive achievement. The arches rose some 39 ft above the Irwell, with embankments half a mile long, 112 ft across the base, 24 ft at the top, and 17 ft high, through which was cut a short tunnel for that turnpike road which had raised objections to the second Act. The correspondent of the *Manchester Mercury* inspected the arches with astonishment, noting that 'not a single drop of water could be perceived to ouze through any of them', but the journalist had missed the main point. It was Barton aqueduct, with all its publicity, which demonstrated clearly that canals need not cling to rivers but could be taken almost anywhere in the land, while its proximity to the boats of the old navigation, struggling across the shallows below, drove the point home with a vengeance. Francis even seems to have changed his motto for the moment from the enigmatic *Sic Donec*, '*Thus until*', to *Perrupit Acheronta Herculeus Labor*, the legend over a contemporary portrait – 'Herculean effort made a way over Acheron' (the river of the dead), in double allusion to the soughs and the struggling Irwell Navigation above which the cut had passed.[9] Clearly the duke's old tutor had been paying a visit to the works.

Barton was an encouraging achievement for all concerned, so early on in the long, hard years of canal construction, and Lowndes was justified in depicting him, looking like a cross between a slightly down-at-heels Gulliver and a conjurer who had produced an aqueduct out of the capacious lining of his pocket – which indeed he had just done. This portrait resembled that of Samuel Egerton at Tatton, his hand stretched over the Venetian waterways. Though Gilbert was obliged to reface the surface of the aqueduct, the rest stood firm until dismembered with great difficulty to make way for the Barton swing aqueduct over the Manchester Ship Canal – one of the seven wonders of British canal engineering. With the aqueduct working, public opinion, so long cynical or downright antagonistic, swung strongly in favour of the duke's plans. Almost overnight the triumvirate became national heroes, and instead of considering Francis a harebrained speculator scarcely out of his 'teens, he was praised as vigorously as he had once been condemned:

> Cotton may boast in his descriptive song,
> Of wonders in the Peak, admir'd so long;
> These works no less a prodigy can boast,
> Where admiration, where description's lost!
> Seen and acknowledg'd by astonish'd crowds,

9 The young duke demonstrates the improved technology of summit-level canals. The artist is clearly imitating the outstretched hand in Samuel Egerton's portrait at Tatton Hall

From underground emerging to the clouds;
Vessels oe'r vessels, water under water,
Bridgewater triumphs – art has conquered nature.[10]

Battle of the broadsides

Once the duke had made up his mind that he could probably afford to run his canal to the Mersey estuary and make a bid for the rich Manchester to Liverpool carrying trade, the triumvirate went into action with their usual formidable efficiency. As an opening barrage the duke issued a broadside pamphlet for distribution to MPs and the public in Manchester, Liverpool and London, enumerating in a short, sharp litany some twenty sins and failings of the Old Navigation, mixed with several encomiums on his new canal idea.[11] He accused the Irwell, among other things, of monopolistic practices, inability to get boats past shallows, of being through bad management a 'precarious, or rather a desperate undertaking', of charging the duke many times more than the agreed 6*d* a ton for carrying material up to Barton and coals on to Salford, of moving boats to obstruct the barges of the Sankey Navigation and the Salford Quay Company so as to engross the conveyance of goods, of deceiving the public over freight rates, and of setting these so high that it became uneconomic to carry coal on such a waterway.[12]

The Old Navigators were not slow in discharging a retaliatory broadside of their own[13] in which they scrupulously placed a reasoned answer alongside most of their opponent's accusations, but it is clear that, even if some of the duke's shots had sailed wide, others had registered, for the Old Navigators had indeed provided a limited, slow and costly service and behaved in a domineering and sometimes monopolistic manner. All the same, they deserved and received a little sympathy, support and consideration, for after struggling for thirty-seven years to build a working waterway they were now faced with that vigorous and formidable competition which they must long have dreaded.

So they issued a second and shorter broadside, stating their case with clarity, claiming that they had spent £19,000 on the waterway, and from 1724 to 1749 had not made a penny's profit on it: indeed, if interest was taken into account 'the whole amounts to *thirty-eight thousand pounds*, and upwards';[14] even so, their paltry return was only $2\frac{1}{2}$ per cent – how then could there be room for a canal? The duke, they claimed, was intending to run his cut very close to their own waterway, and ruin, they felt, stared them in the face; when the new canal took water from the feeders it must lower their own supply.[15] They had boats lying idle for lack of trade – how then could there be room for a new transport system? They would, they threatened, meet this challenge by cutting their rates to 6*d* a ton.

The duke's return broadside made, initially, much the same case as his parliamentary petition presented on 14 November 1761[16] in which he sought permission to extend his canal from Longford Bridge to the Hempstones on the tidal estuary about eight miles downstream from Warrington Bridge. He offered the Commons the happy news that his coal canal to Manchester and his 'subterraneous sough' were both getting along very nicely, with the underground waterway carrying boats of six or seven tons burthen. He had, he claimed, taken his surveys, though Brindley did not actually get through to the estuary until 18 November, and then set off for London the next day to help with the evidence[17]

The duke's most astringent argument, which won him powerful support against an increasingly tough alliance of landowners and Old Navigators, was that the Irwell was still 'very imperfect, expensive and precarious', so that boats could not struggle up to the lock at Warrington Bridge except at very high tides, while goods from Manchester to Liverpool were diverted to the roads at charges of 30*s*–40*s* a ton. The Irwell was charging 12*s* a ton for the same service, but the duke was willing to cut this to 6*s*, and so halve the Manchester–Liverpool freight rate – as tempting a bait as any merchant could ask for. A particularly telling point was that the Old Navigators had neglected to serve the counties by failing to build a single wharf 'between *Manchester* and *Warrington Bridge*, which is upwards of 26 Miles . . .'.[18] He insisted that he had no intention of taking any water from the Mersey and Irwell because those rivers lay below the level of his cut: 'The Canal is to be chiefly supplied with two

Streams of Water flowing from Springs cut by the Duke and his Ancestors in the Bowels of their Estate, by driving up Soughs to drain the Mines, and the Waters from thence are still increasing by the Progress of the Works, the deepest of those Soughs being now carrying on with great Expedition; these waters, it is apprehended, the Duke may with great Propriety call his own', and he claimed to have enough already to fill eighteen locks in twenty-four hours – far more than the traffic then warranted. If any slight loss to the rivers could be proved, he was willing to make 'full and ample Satisfaction'.[19]

The vast expenditure and minimal return of the Old Navigation was, the duke claimed, no argument for preserving it, but rather a sure indication of its total inadequacy: 'and is the Public', he demanded scornfully, 'to be denied a more expeditious, safe, and in every respect a better' canal because the hopes of the public and the intentions of Parliament had been largely baffled by the Irwell Company?[20] Turning his fire on to the turnpike road adventurers, he contrasted them unfavourably even with the undertakers of the river navigations, arguing that the new canals could obviate the wasteful custom of passing Acts for river navigations and turnpike roads to run alongside each other when neither could carry goods very efficiently. He painted for Parliament and public alike that vision of a national system of arterial canals which his own example and tenacity would one day serve to create: 'This mode of Navigation is new of its Kind . . . and may, if the Completion thereof is allowed, be the means of introducing into many other trading parts of the Kingdom a more easy, Cheap, and expeditious Conveyance than can otherwise be obtained.'[21] Finally he argued that any new scheme invariably met with *local* opposition, but added flatteringly that Parliament had a splendid reputation for a wider view which fostered 'every attempt where Public Utility was the Object'. The duke's recurrent insistence on the virtues of utility implies that he was acquainted with David Hume, the philosopher and historian who influenced Bentham and the Utilitarians.

While these broadsides were being exchanged, the battle was raging no less fiercely in the Commons. By 21 January 1762 the Bill was in committee, with counsel examining John Bradshaw, a

Manchester boatman who had navigated vessels on the Irwell for twenty-five years.[22] Though Bradshaw maintained that the public was well pleased with the old waterway, he was forced to admit that, even when loaded with a light cargo like timber, his record voyage from Liverpool to Manchester was a full week, and, yes, Hollin Ferry was very shallow in summer. Gradually it emerged that barges had to cut their tonnage from thirty-five in the winter to eighteen in the summer owing to shortage of water – powerful support for the canal lobby – and as another boatman, Thomas Unsworth, pointed out, the flats drew at least 3 ft, so the skipper had to estimate the likely depth of the navigation, and load accordingly, before setting sail.[23] These flats derived their name from flat-bottomed sailing barges – as handsome and distinctive a breed as the Thames barge itself.

On the other hand Joshua Taylor, a miller with, he claimed, no interest in the old navigation, insisted that the Mersey and Irwell could carry all and more than the existing trade, while the duke's cut would, he believed, be of no service to the merchants of Manchester.[24] Under cross-examination some witnesses admitted that the river locks often had to be opened to flash boats over the shallows, but others argued forcibly that an arterial canal would freeze up in winter and halt the traffic, while the Irwell ran fast and free. Yet it was generally admitted that the flats regularly lay four or five days at Warrington for lack of water in the summer, and 'had to wait at Liverpool 2–3 weeks for loading, and took 2 weeks about, in coming up the river'.[25]

Meanwhile Brindley had not been kicking his heels long in London, but was riding as hard as ever in the canal cause. On 7 January 1762 he met John Smeaton, the most brilliant and distinguished engineer of that age, to measure the sough water supply at Worsley. The reason was that, while the Commons might permit a new competing canal which would greatly improve the area's transport service to the public, it would never tolerate the theft of water from streams feeding the old navigation, so the duke was obliged to prove that he could obtain a very adequate supply from his soughs. In fact he was always a little short as traffic increased beyond the wildest expectations, and towards the end of the century was still personally supervising the construction of large reservoirs higher up on his estate.[26]

Smeaton's evidence

If the laurels for the first Act should be awarded to the duke and Gilbert, and for the second to Brindley with Gilbert, the third Act was unquestionably John Smeaton's.[27] He had been elected a Fellow of the Royal Society in 1753, and in 1755 had designed his famous Eddystone lighthouse out of dovetailed stones; while Brindley was achieving distinction locally, Smeaton was heing honoured on a national scale. Gilbert opened the case for the duke's petition, but it was Smeaton's highly involved but authoritative technical evidence on water supply, combined with an uncompromising conviction of the superiority of canals over navigations, that carried the day. First he dazzled the committee with mathematical calculations, and stated that he had measured the Irwell supply and found it flowing adequately; then he predicted that, far from losing by the sough water, the old navigation was more likely to profit by it, for as the tunnels were driven into the mines the volume would increase and the overflow must benefit the river below. 'Those,' he added 'who have experienced Canals would chuse to Navigate in them when they can.'[28] Asked whether water diverted to the duke's cut from Worsley Brook might make a difference to the depth of the river, he replied that his calculations suggested only 'a difference of 100th part of an inch'.[29] He had also allowed for evaporation – an important factor in water supply.

Under stringent cross-examination on the loss of water through leaking locks he replied, 'It depends upon the Construction. It is possible to make them with little Leakage.'[30] This was a modest statement, for Smeaton was himself a pioneer of improved locks. Mitre-gated locks had been known in England for some time and consisted of two doors meeting at an apex so that the weight of water thrust them together – they are still used today. Smeaton had mentioned in parliamentary evidence for the Calder and Hebble Navigation in 1758 that he could build 'locks better than most of the locks he had seen on any river in England', adding gently that there was no reason for any of them to leak excessively, and most of the MPs would have known of his achievement in that field already.[31] As the cross-examinatin continued Smeaton guarded himself against questions irrelevant to water supply, reminding counsel that he was not the engineer to be employed.

After Smeaton's profoundly impressive and professional but dryly reserved evidence, the committee must have looked forward with relief to another performance by Brindley. The duke had kitted out his consulting engineer in a fine new coat and waistcoat of broadcloth, with new breeches, shoes and buckles, and a new penknife in case he decided on further visual aids. On 17 January 1762 John and Lydia Gilbert took him to see David Garrick in Shakespeare's *Richard III*, the play in which Garrick had first made his reputation in 1741, and from the moment the twisted hunchback sidled on to the stage he held the audience spellbound:

> Plots have I laid, inductions dangerous
> By drunken prophecies, libels and dreams . . .

Brindley was an exceptionally sensitive man, and the smoking falchions and bloody murder set his mind on edge and his imagination working overtime. He retired to bed until Sunday, and went to Mattins at St Mary's Church to recuperate. Meanwhile Francis stayed with his mother at her fine new house off Piccadilly, so there had been a long overdue family reconciliation. The sight of the son she had once branded as an imbecile in fine health and working on the most controversial issue of the day, surrounded by his brilliant staff, must have been a salutary lesson for Rachel Russell.

Brindley's evidence

Called before the committee on 28 January 1762, Brindley, his mind still perhaps a little clouded by Shakespearean misconceptions of medieval tyranny, took longer than usual to warm to his subject. He claimed to have been 'employed by the Duke of Bridgewater ever since the work begun', and at Barton Bridge two summers.[32] The context here suggests that Brindley meant ever since the survey which followed Taylor's Trent and Mersey, but not on the Salford Canal. At Barton he had noticed the Old Navigators throwing out tackle to haul their flats over the shallows: they used men and not animals for bow-hauling, and even then they needed one or two flashes from the locks to help them along. Waxing a little more eloquent, he insisted that the new canal would bring boats from the Hempstones to Manchester in one day and save eight or nine miles – 'can't say,' he added gruffly, 'that we have any Benefit from the Old Navigation.'[33] On 13 December 1761 the duke, with his parliamentary foresight, had sent Brindley to Warrington Bridge to count traffic into Manchester and so confound the opposition theory that there was not enough of it to warrant a new canal – there he had seen 127 carts, three waggons and 385 loaded horses pass by on a single Saturday.[34] Cross-examined on the water problem, Brindley got back into his old form as an orator, assuring them that he had 'as much experience about defending water as anyone', which he preceded with the startling statement: 'I dare undertake to make a lock that will lose little – I make upon a new Principle . . . I can make a lock that won't lose a gallon of water in an Hour.'[35]

'How long,' he was asked, 'do locks last as made at present?'

The reply came: 'Perhaps a twelve month, and perhaps not two months.'

'How long will the lock last you propose to make?'

'Ten years or more without repairing in that state of perfection as not to lose more than one gallon in an hour – I don't mean in running water,' he added, to drive the point home.[36]

It must have been an electrifying moment in the Commons when this announcement was made, but it is hard to assess how much of it was rhetorical exaggeration, to which Brindley was rather prone – it is difficult to believe that even the crudest lock could last less than a year. Brindley had carried out important work on dams in Yorkshire and at Trentham, and had great experience of mill pools and races, but there were as yet no locks on the Bridgewater Canal, or evidence to suggest that he had yet built one, so it seems likely that when he had met him a fortnight earlier to measure the duke's sough water Smeaton had, with his usual generosity, told his colleague the secrets of his recent improvements in lock construction.[37] Brindley then outlined the course of the canal through Cheshire enthusiastically, saying that some eight miles had been cut already towards Manchester, and that the duke was using about twenty boats. As consulting engineer he was convinced that the new canal would not take water from the feeders supplying the old navigation – 'we go over them all,' he insisted, so it became clear that, initially, the Bridgewater Canal resembled Henry Berry's Sankey Navigation, or the Bereguardo in Italy, drawing headwaters chiefly, and largely independent of feeders – the duke and Gilbert probably used both as models. As it was extended, feeders from the Medlock and the Rochdale canal added supplies.

In further evidence the opposition from the Cheshire landowners grew extremely powerful – Samuel Egerton had friends, relations and supporters among his neighbours around Tatton, but the sanctity of private property was at stake on a national scale. The proprietors of eleven miles of land along the new course had signed a petition protesting vehemently that it would divide their fields and force cattle to be driven miles round by bridges. Sir Richard Brooke of Norton Priory near Runcorn owned the second largest stretch after Lady Stamford, and his Manchester cousins had held the third largest holding on the Castlefield line, so they were united in their determination to fight every inch of the way. The parsons, it was pleaded, would have to travel two miles more to collect their tithes; blacksmiths, who rode far and frequently, would fare worse; while land next to the Sankey, which had been 'cut on the same plan as the intended one', was waterlogged and reduced to half its original value. There should, it was argued, be a bridge built for every single farmer, but why bother? A small expenditure could make the old navigation into a really efficient waterway – a point well parried by the duke's counsel, who emphasised that fifty earlier shareholders had forfeited their holdings rather than pay further calls on such an uncertain enterprise.

The duke directed and supervised all his affairs, but he took a particularly active part in the 1762 Bill, sending out a total of 750 letters to Members, sympathisers and waverers. His chief opponent was Lord Strange, the courtesy title until succession to the peerage of the sons of the Stanleys, Earls of Derby and former Kings of Man, who were major Lancashire landowners in the hundred of West Derby, near Liverpool, and were endowed with ancient Cheshire lands at Storeton, and the hereditary Forestorship of the Wirral. The Stanleys also had special and ancient links with Salford, where the Byroms were among the main supporters of the Mersey and Irwell Navigation, and were the Stewards of Macclesfield Forest.[38] Strange, an able politician, mustered a formidable battery of predominantly Tory interests – the Old Navigators, the packhorse and coasting trade men, and an impressive array of angry squires who saw private property as something sacrosanct. These stood opposed to the duke's mainly Whig interests, for Gower and Bedford were of that persuasion, though Francis was more nearly attached to the court party, his area of special interest being the canal vote, over which he exercised a lifelong domination. By February 1762 Brindley was noting in his diary that Strange's supporters were gaining ground against the duke, but on the critical vote the canal lobby won by 127 votes to 98. Brindley also recorded that the leader of the opposition was sick with grief, but Strange fought on through several hopeless divisions. Anyone reading the surviving evidence impartially at this distance of time must be inclined to consider that this was a just decision, and that Francis had won his case more on its merits than on party or financial influence. The Mersey and Irwell had provided a slow, expensive and incompetent service at times, fish garths over which there had been lawsuits still blocked the navigation, and it had treated its customers in a high-handed and monopolistic fashion: 'blinded and infatuated with self interest' John Hart, a Warrington merchant labelled them, and though he may not have been an entirely disinterested observer the majority did share his view that posterity would 'experience the beneficial effects and speak with honour of the noble patriot' who was launching the canal idea.

The duke's third Act received the royal assent on 24 March 1762, but Francis was certainly not the canal megalomaniac that some writers have so luridly painted. In the same year he found time to heed Pitt's prediction that Canada would be won on the banks of the Elbe, and voted against the withdrawal of British troops from the Anglo-German army led by Ferdinand of Brunswick. In 1763 George

Grenville succeeded Bute and, believing that North America should have a standing army and contribute to its upkeep, imposed the controversial Stamp Act in 1765. When Grenville retired Rockingham succeeded him and the Act was repealed, but the duke voted against this motion. From being among the most insignificant rulers of his country he had become one of the most controversial, but here and there, and particularly in the House of Lords, where men felt the pulse of the nation and understood more clearly what was happening to the eighteenth-century, he was gaining respect as one of a small but growing band of outgoing men who had the employment and welfare of industrial people much at heart. After the third Act that inspiring toast of the canal age would be heard more often where men gathered together and raised their glasses – 'Success to Navigation'.

Notes to this chapter are on p. 187

8

The underground canals

Francis Egerton was, perhaps, the first English landowner to open his grounds to the public[1] for a fixed charge. Accommodation for his growing staff was such a problem that he built the imposing Brick Hall at Worsley. His heirs in the nineteenth-century built a third hall a little to the south, a very large stately home overlooking the canal. The Brick Hall was the central office of his mining and waterway enterprises, where the adventurous visitor could obtain tickets for Gilbert's underground canal system – the first example of its kind in Europe, according to the visiting mining engineers Fournel and Dyèvre.[2] The duke opened his soughs to selected visitors because he needed support for his canal Acts, and was anxious to publicise the value and novelty of his enterprises – it also brought in a little more badly needed money. By the time working ceased in 1887 the underground canal system had covered no less than forty-six miles, with an additional half mile of navigable sough at Boothstown which was built in 1822 and linked a mine at Mosley Common with the open-air canal. Later in the century it was said that a man could go down a shaft in the heart of Lancashire without needing to touch land again until he reached America.

The canal idea which the triumvirate was busily engaged in publicising was important and overdue, but it had long been known on the Continent. The underground canal system was entirely different – it was an original and uniquely English contribution to the industrial revolution and to the science of engineering, and, though initially expensive, saved so much winding of coal to the surface, and so much costly draining of mine water, that it proved in the long term an immensely profitable investment. It was therefore imitated in many places – by the inventor himself at Harecastle tunnel and his Alston Moor silver mines; at Lilleshall in Salop for Earl Gower, by foreign engineers at Fuchsgrube in Lower Silesia, and in limestone mines at Dudley to the west of Birmingham, as well as in several of Gilbert's mining projects in Derbyshire. Bernard Shaw accused the British of neglecting their geniuses, and Gilbert and the duke can be taken as cases in point. The Worsley sough system alone has been justly described as 'fantastic, certainly it is unique and it must rank as an outstanding achievement in coal mining history'.[3]

The coal ransom

In Lewis's *Manchester Directory* of 1788 there is a vivid description of the miners' combine which preceded the duke's operations. 'Their custom was to get coals and keep them in the pit, while they went up and

10 The quarry and subterranean tunnel at Worsley in 1766

idled away their time, or were drinking at the ale-house, till their demand was complied with of a gratuity, which they levied as a tax on the carters, who were there waiting for coals, to the great loss of their time and ruin of their teams.' Only one other mine – at Newton – is mentioned as a rival to the duke's in the immediate vicinity of Manchester at that time, and it had suffered severely from flooding until the installation of a steam engine, about 1759, stepped up its production again.

It was one of the duke's favourite maxims that, if it was to prosper, a canal needed 'coals at the heels of it'. Heel is the operative word, for it was the carrying trade in general merchandise, the imports and exports of growing Manchester, which eventually brought the canal its highest returns, but coal sales did not fluctuate so much as the general trade, which depended more on the national economy – a canal did need coal at its heels to give it stability. While Brindley was coping tolerably well with Barton aqueduct the duke was running into grave financial difficulties[4] by spending far beyond his means. Claims that he built his canal relatively cheaply have been advanced, but they entirely overlook his land and mining purchases, and if we glance at a few of his outgoings from 1762 to 1765 we can see the reason: an acre of land at the Liverpool dock had cost him a reputed £40,000,[5] while Wardley Hall and other mining rights bought from Lady Penelope Cholmondeley in 1760 may not have been much less.[6] George Lloyd's Hulme Hall estate and Humphrey Trafford's lands on the Castlefield route added at least another £13,000 to the bill; by 1765 his wages and salaries were running high, and the canal debt stood at £60,879 and was growing daily.[7] The Brick Hall must have cost a moderate fortune, and flats, of which he soon had twenty, £7,000 in ten years for those built at Bangor on Dee alone; not all the land purchased seems to have been entered into the accounts, implying that Francis sometimes paid from other resources,[8] while the growing docks and warehouses at Castlefield, Manchester, with the soughs, stabling and offices must have added substantial sums, for where the duke built, he built to last.

Francis was not always the wise business man that posterity has painted him – he learned by a hard process and certainly overreached his resources at this time, being forced to mortgage the canal works to his family banker, Sir Francis Child, for £25,000, a sum which vanished rapidly into his voracious waterway. Perhaps this urgent and even extravagant spending was due to a feeling that time was short, after his favourite sister, Louisa, Lady Gower, had died in 1761.

The Egertons probably had only two mines still working by 1759 on Walkden Moor, so further substantial expenditure was needed to purchase mineral rights from other landowners deep into the Lancashire coalfield where the soughs were penetrating; the normal terms were a duty of one-eighth value of the annual output – a condition which ate deeply into his ceiling of 3½*d* a cwt to which he confined his sales until 1793. By this process he gradually acquired the coal rights in most of the district.[9] The Manchester and Salford area was soon his main market, and as the canals progressed he was selling around 400 tons a week there in 1780, and 130 tons at Broadheath, with much smaller quantities off the canal; in that year he made a gross profit of £11,000 – cannel, the hot, fast-burning, high quality coal, earning double the statutory price, but he had little of it.[10] Despite these problems Francis does not seem to have ignored à Kempis's injunction that we are sent 'to serve and not to govern'. The agents at all his depots had strict instructions when coal was scarce, as it often was, to give priority to the poor people who came with barrows and aprons, while the rich with their carts and waggons were sent away empty until supplies picked up again. Very different was the attitude of the Liverpool dealers, who sometimes gave bad measure on coal coming mainly from the Sankey Navigation, and 'seem to have been reluctant to sell in the small quantities which the ordinary householders desired'.[11] Francis had too many other commitments to be at Worsley all the year round, but his visits were frequent. Dressed in his famous old brown suit which – that last abasement for a once fashionable beau – was made by the colliers' tailor at Worsley, he would wander round his works, watching happily as poor people came to buy coal at his reduced prices, helping to hoist a sack occasionally, or meditating on new methods for relieving that abject poverty which lay all around his wharves. A newly engaged lad saw him apparently standing idle one day and shouted, 'Here, fellow, gie us a hoist up,' with a sack he was resting, and when his employer had done so, commented with engaging candour, 'Well, tha's a big fellow, but tha's a lazy un' – a remark which was far from true. Francis's careless disregard of fine clothes was the despair of his one remaining servant, Aubrey, who would constantly remonstrate that dukes ought to go around properly dressed. When hard pressed on this point in London the duke would sagely reply that it did not matter there because nobody knew who he was, but when nagged in the same cause on his estates he would innocently insist that it could not matter there either, since everyone knew who he was.[12]

Labour supply

By April 1762 the duke's mines were beginning to take up the slack in local unemployment with an urgent advertisement in the *Manchester Mercury* for 'Any number of Sober, Diligent Colliers to whom all reasonable encouragement and accommodation will be given upon Application to Mr Gilbert at Worsley',[13] followed two years later by one for 'Honest, Industrious Colliers, Sinkers and Soughers, that have been used to work in coal Mettles', but these were not enough, and he was soon sending agents to recruit in North Wales and farther afield. A particularly close relationship sprang up between the duke and his miners, and though Gilbert had to be a little stricter he was highly respected for his brilliance as an engineer, so the duke acquired a reputation as a model employer for those times, when life was sometimes desperately hard for working people. All the staff and

W A N T E D,

At his Grace the Duke of *Bridgewater's* Colliery, at *Worſley*, near *Mancheſter*, ſome Honeſt Induſtrious Colliers, Sinkers, and Soughers, that have been uſed to Work in Coal Mettles; where all reaſonable Encouragement will be given them, by applying to *John Gilbert*, at *Worſley* aforeſaid.

11 Advertisement for miners

labourers were obliged to belong to the estate's sick club with a quarterly subscription of 2*s* which provided a variety of benefits, but as the century wore on illness increased, especially as the miners grew older; the club became bankrupt and the duke had to bail it out.[14] A visiting clergyman, the Rev. T. Gisborne, noted that the Worsley colliers responded well to the improved conditions of employment – 'a neighbouring magistrate informs me that he hears few complaints from either side and that the colliers are more moral than the weavers . . .'.[15] By 1780 the Church was taking a more serious interest in general education when Robert Raikes and the Rev. Thomas Stock started the first Sunday school. Though such schools may not seem very impressive in an age of national education, they were a major innovation, included the teaching of reading and writing, and spread quickly to Worsley, where Francis gave his land free of charge for all new buildings: 'As soon as your engagements will permit I wo'd thank you to order the ground to be set out for the Hindley Sunday School Houses, if you please,' wrote the Rev. Mr Peters to John Gilbert in August 1787.[16] By that time the duke was surmounting the worst of his financial problems, but it was this kind of unhesitating generosity which drew an answering response from his people – in 1783 there were 207 miners and 124 drawers employed at the Worsley pits, but this figure probably increased substantially over the next twenty years.[17] The 'bord and pillar' system which left columns of coal to support the roof was used in the duke's mines, but there were some places where long-wall working pushed the tunnels forward in a continuous line, the drift being filled with stone and slack to support the roof.

For all that he was a generous employer, Francis kept a close eye on his men when he was at Worsley, since Gilbert was often away on his multitudinous enterprises. On one occasion he stopped a collier who was late to work and ticked him off, but the man gave the excellent reason that his wife had had twins in the night – a doubled worry in a growing family. 'Ah well,' said the duke piously, 'we must accept what the Lord sends us.' 'Aye,' said the miner, 'but I notice he sends all the babies to our house, and all the brass to your'n!' – a home truth which made the duke laugh, and earned him a tip of a guinea. Others with less convincing excuses were fined half a crown by Gilbert and his staff, who paid this money to the building of a row of new houses for the miners which was ruefully dubbed Half-crown Row. The estate ran its own shops, where the truck system worked well because the tradesmen were tenants-at-will. A close eye was kept on them, and they dared not overcharge.[18] The canal was indeed important, but it was the coal that would reduce local poverty, and there the duke and Gilbert seem to have concentrated their industrial relations. The account books are filled with records of their special treatment of the colliers, often ordered by the duke in person when he 'rewarded them for any extra or particularly arduous work with ale and food, and also with bonus payments'.[19] It is therefore hard to excuse the chimera of his meanness, so long reiterated.[20]

12 Woman wearing belt and chain for hauling in the duke's mines (*by courtesy of Frank Mullineux*)

For all that these miners were well paid compared with other classes, conditions even in the Worsley collieries were appalling by modern standards. They worked an average of twelve to fifteen hours a day for six days a week, major feast days excepted, for about 8*s* a week, while the women and children earned from 4*s* to 1*s*. The duke's pits evolved a system of payment on a fractional basis: a boy under ten years – and quite a few were employed – was counted as one-eighth of a man, at ten years he became a quarter, at thirteen he was three-eighths and at fifteen he was a half; the assessment of a fully grown woman would scarcely have pleased the advocates of Women's Lib., as she counted as only half a man, though one miner said that he always preferred women as drawers because they were more obedient and kept better time, though they fought and shrieked a lot.[21] Posterity has, perhaps fortunately, been left no record of what the women thought of the men. Often a boy and a girl would work together, and one lad of seven years who equalled one-eighth and a girl of fifteen who was a half had done about seventeen and a half miles of drawing between them in a day, though this was exceptional, the seams often being low and the coal cumbersome. They used heavy sledges shod with iron runners, the child in front dragging a chain which ran between her legs, while the one behind pushed uphill, or braked down. There is no record of Worsley using the horrifyingly dangerous method of carrying coal up rickety ladders, which crippled so many, but as the soughs progressed the men usually went to and from work by laddered shafts. Mr S. Greenhalgh, who reladdered Scott Meadow workings in 1913, recalled 'the great effort required to climb those ladders – very nearly the height of Blackpool Tower. It makes one appreciate how arduous the work on the canals must have been. . .'.[22]

Technically the miners were engaged on contracts which could not be broken by law for a year, but in fact the duke's men tended to stay with him anyway.

The mine canals

It was the curious repetition of the faults and the angle at which they lay that chiefly influenced Gilbert and the duke in their decision to use a canal system when the first survey was made, combined with the availability of water and the accessibility of the workings. It was therefore not feasible to use navigable soughs for the majority of other mines.[23] Two tunnels were driven in from the Delph about thirty yards apart and some 82 ft above sea level; after 500 yards or so they joined at a spot called the Water Meetings (not to be confused with Water Meeting, near Stretford), the overall height being 8 ft with some 4 ft of water below and a breadth of 10 ft, with the first seam lying about 750 yards from the Delph.[24] Surface drilling was carried out extensively all through the duke's time to see where the best seams lay – probably by the percussive process with hand turning which was long used on shallower bores.[25] The duke found this prospecting particularly exciting and awaited the final results impatiently, watching as the bits delved nearer to depths where new discoveries could be expected. His presence embarrassed the drillers, who, out of respect for their employer, did not like to stop work when the stable clock struck twelve, until eventually an old hand explained the problem to him. After that Francis always walked away punctually, but on returning at one o'clock often found that he was the only person there. When he asked the reason they produced the ingenious excuse that though they could hear twelve strokes they often missed the single chime of one. The duke magnanimously undertook to aid their hearing with a special mechanism which made the clock strike thirteen. This clock is now in the church spire at Worsley, and was stopped during Hitler's war lest German invaders, drilled with Teutonic efficiency in the minutiae of local history, might learn their location from it. Another was installed over the Bridgewater office at Walkden.[26]

The work on this extraordinary underground canal system was inevitably slow. The reporter from the *Gentleman's Magazine* in 1766 bought his ticket, was issued with candles, and boarded a 'starvationer'. 'Through this passage you proceed, towing the boat on each hand by a rail . . . before you come to the coal works; then the passage divides,' one going on to the coal seams 300 yards further, and the other about 300 yards to the left – and he noted that it was planned to extend them.[27] Ventilation funnels ran up from niches either side, aided by giant bellows fixed at the Delph, pumped by a diverted stream falling into a funnel. This injected a jet of air and water into the entrance for cooling and circulation, but the soughs made these mines cooler than most – though probably built by John Royle, one of the duke's mining engineers, this invention is traditionally attributed to Brindley.[28] The drawers filled small waggons which tipped into starvationers holding some eight tons each at that time, and the reporter stated that there were fifty of them already working; by 1842 this had increased to 150 ten-ton M boats and 100 two-tonners called tub boats.[29] Winding coal to the surface up the traditional pits was expensive, and steel ropes were not introduced to Britain until 1835, so the economy of Gilbert's system can be seen from a record established by one of his lads, who drew twenty-one starvationers carrying eight tons each out to the Delph. This makes a total of 168 tons hauled by one man, though much aided

13 A junction in the soughs: the underground canal system at Worsley, in Lancashire (*by courtesy of Frank Mullineux*)

by a clever system of raising the clows at the Delph to flash the loaded boats gently down the tunnel.

Gilbert was fortunate enough to save his employer considerable sums of money by discovering lime deposits on the estate – vital to making mortar in the days before cement. Immense areas of the sough canals were roofed with hand-made bricks, and there was a great need for them at the Brick Hall, at the warehouses and in many works; this Sutton lime from Chat Moss was not always the right quality and some had to be imported from Bedford, an area of nearby Leigh, though there was invariably a little left over from the works for sale to the public.

The immensity of the whole operation can be seen from the dimensions of some of the soughs – sometimes 16 ft wide with junctions spanning 26 ft, so that the mind boggles at the achievement of the men who hacked those forty-six miles, mostly with primitive tools, or bricked the low arches yard by yard. Work was by contract, and only about six men could operate on a heading at any one time. Yet the picture that emerges at the end is by no means one of slave labour grudgingly given, but rather of men, women and possibly children comparatively happy, paid as an elite corps under the duke's special consideration, often dedicated to a uniquely new and exciting project, and giving good measure to an employer who knew most of them personally.

Nor was any waste allowed in these undertakings. 'The Duke, like a good chemist, has made the refuse of one work construct the material parts of another . . .' wrote the reporter, noting how the sough stone went to Worsley Yard for shaping by the masons and on to the open canal to build works like Barton aqueduct, the bridges and the culverts, while clay taken from shafts went to puddle the canal – 'In short, the noblest design has been conducted with the strictest economy'.[30]

Nowadays people seldom travel far to see a new engineering project, but the eighteenth-century offered few alternative distractions. Once the *Gentleman's Magazine* had publicised the soughs people of fashion crowded to see them. Clutching the candles issued by the boatmen, James Bury, perhaps, or James Bromilow, they would crouch in the narrow starvationers in some trepidation, comparing their trip with a visit to the underworld, though one of them noted that he met with more civility than Homer or Virgil had described in their visions of Hades. 'Should your spirits sink,' one visitor added, 'the company are ever ready to assist you with a glass of wine.' The great doors at the Delph closed silently to prevent draughts which could extinguish candles that only tinder could relight, though he claimed that these tapers tended to make the darkness more visible.

> But this dismal gloom is rendered still more awful by the solemn echo of this subterranean lake, which returns various and discordant sounds. At one moment you are struck by the grating noise of engines, which by a curious contrivance let down the coals into the boats. At another you hear the shock of an explosion, occasioned by blowing up the hard rock, which will not yield to any other force but that of gunpowder; immediately after which, perhaps, your ears are greeted by the songs of either sex, who thus beguile their labours in the mine.
>
> When you reach the head of the mine a new scene opens to your view. You behold men and women, almost in their primitive state of nature, toiling in different capacities by the glimmering of a dim taper, some digging the jetty ore out of the sides of the earth, some loading it into little waggons for that purpose, others drawing the waggons to the boats.[31]

Lord Stamford and other local landowners held house parties for those visiting the works and Barton aqueduct; in 1772 Thomas Pilkington, captain of the passenger boats, was tipped generously for looking after a tourist described in the accounts with peculiarly English scepticism as 'reported to be a Poland Prince'. In 1778 Christian VI of Denmark, who had married George III's sister, Princess Caroline Matilda, made what was supposed to be a surprise visit to Salford, and put up at the Bull Hotel where Gilbert had received his inspiration. Accompanied by Edward Byrom and other local dignitaries, his party embarked at Castlefield, Manchester, crossed Barton and went up the sough in starvationers; after which he said how much he had admired the greatness of these undertakings, and distributed handsome tips. Gilbert seldom allowed the boatmen to retain much of this, but put it towards the sick club.

John Gilbert was the first man to suggest blasting with gunpowder for rock salt at Marston, near Northwich in Cheshire,[32] and he was certainly using it in the coal mines by 1759 – probably earlier. Meanwhile the debt for financing all these activities was growing at a phenomenal rate, and continued to increase until 1787, when Thomas Kent, pressing hard on his quill, wrote, 'debt decreased this year £434–7–7*d*'.[33] The duke felt that the worst was over by 1783, when he once again showed his special consideration by ordering a bonus of 21*s* for each collier, and 10*s* 6*d* for every drawer who had been in his service a year or more.[34] By that time Francis was certain that there were vast reserves of fuel in his mines. Originally there were only the twin navigable soughs running in from the Delph, but when the Gilbert system had proved its worth another underground canal was drilled on a higher level than the Delph one. It followed the ancient adit in use since Scroop's day, starting a little north of Dixon Green and emerging near the present St Mark's Church at Worsley; this the duke linked to his main line by an inclined plane which will be described later. A third level was begun – also for boats, and probably during the duke's time, 56 yards below the main sough, and the fourth and lowest canal lay 83 yards below the main line. Both the lower canals were working in 1822, and the deepest underground one contained good-quality coking coal, though only in narrow seams.[35] The boats were not propelled by rings or ropes for very long in the Worsley mines, because of the lack of overhead room as silting increased, the halers preferring to leg the boats by lying on their backs and walking their feet along the brick arches, a process still used well within living memory, like the unofficial title of 'Admiral' awarded to the master haler. The air was none too good in these workings and gas was something of a problem; in one place above the inclined plane gas jetted from the wall like a burner and was kept alight as a landmark for the halers, who often worked in the dark, and in order to light candles in that age of flint and tinder.[36]

When the future President of the Royal Society, (Sir) Joseph Banks, visited Worsley mines in 1766–67 he noted that seams were already being opened below the main workings. 'In the lowest part of these works the duke is obligd to lift water, which has given opportunity for shewing an Engine for that purpose, of an intirely new construction, whose very powers & principles have never been thought of.'[37] The inventor was Ashton Tonge, the duke's master miner, and it worked by a head of water plunging down wooden pipes to drive a piston. This powered two pumps which raised the contents of the flooded levels to areas where they could drain away. There was a 'clack' or wooden gear, to lay on or disengage the power at will, 'much in the manner of a fire Engine', Banks commented.[38]

Where the duke's pits were not linked by canal another ingenious water-driven engine drew coal to the surface. Heavy drums of water acted as weights to counterbalance the coal as it came up, and then, as the drums were emptied, they lightened to the surface for refilling and lowered containers for more coal – a process typical of the duke's and Gilbert's skill and economy.[39] The larger M boats travelled to Castlefield, Manchester only – they did not deliver to other depots, as the main sales were always there. By 1842 the economy and efficiency of the underground canal system was paying mighty dividends, with output topping 300,000 tons of coal a year, but by then, Wigan often outsold the Worsley product in Manchester because of its finer quality,[40] while mine working in the Salford area soon met more of the local demand.[41]

From 1760 to 1786 the colliers were generally paid fortnightly and after that usually every month, but as the soughs delved deeper a dispute arose about the most practical place for this operation. It was easiest for the accountants to issue the money at 'Shavingham' (Shaving Lane), where the men emerged from the shafts, but some of the weaker vessels arrived home at Worsley roaring drunk after spending far too much of their families' sustenance at pot houses on the moor. One can therefore see the strong arm of the miners' wives behind a poignant petition which the more sober elements presented to Gilbert in May 1772, begging that payment should be made at Worsley Mills as in the past, 'there being at the latter place and near thereto several convenient places for buying bread, flour, Cheese, Butter, Bacon and Grocery Goods . . . and since they have been necessitated to travel further to procure such commodities, and often before they can compass the same, a part of their money has been otherwise exhausted for what was less usefull in their families, or of no use at all . . .'.[42] This interesting document in industrial relations shows the growth of a sense of corporate social responsibility and concern, probably reinforced by increasing Methodism.

In 1760 the duke's two coal mines on Walkden Moor made a combined profit of only £406, of which £80 came from the smaller Cannel pit. By 1769 the gross profit had risen to £1,420 before deductions, but beyond the close of the century, when the price limit had lapsed, it was running, in 1803, to £24,300 gross.[43] That the duke and Gilbert with their many assistants, accountants and foremen should have devised a system for moving millions of tons of coal for innumerable miles in boats, on four different levels, hundreds of feet deep under the heart of Lancashire – a system which lasted for well over a century – is therefore one of the most extraordinary and least acknowledged miracles of the industrial revolution, seldom granted even a passing mention in any of the textbooks. Nor should it be forgotten that the poorer people received preferential treatment in buying the fuel which this outstanding engineering achievement had made available, by Act of Parliament, for their use.

One of the historians who appreciated something of this achievement was J. H. Clapham. 'From the first,' he wrote, 'coal transport had been a dominant factor in the canal movement. The fuel famine of the eighteenth century would have stopped the growth not solely of industry but of population, in many districts, had no means been devised of overcoming it. The Duke of Bridgewater was a coal owner and his canal had halved the price of coal in Manchester.' Eight years later the first section of the old Birmingham Canal, modelled on the Duke's, achieved much the same for Birmingham.[44] The age of coal transport which Francis inaugurated was a major technical advance on the ancient practice of using for fuel such desperate expedients as the dried dung of cattle.[45]

Notes to this chapter are on p. 188

9

The cut goes west

If the duke's underground canals have received a poorer press than such an achievement deserves, his open-air waterway has been to some extent misunderstood. We have seen that he had determined from the start to make his canal and mining enterprise an extension of his Worsley estate, financed by private loans, and not the type of limited company more readily understood by modern townsmen; we have also seen that he owned eleven other estates, with Ashridge in Hertfordshire as their administrative centre. It follows that the managerial structure resembled a kingdom in miniature, governed by the duke, who certainly made all the major decisions – but he could not always be there in person, since he was busy planning strategy, administering other estates, supervising his parliamentary boroughs, Acts and patronage, and dealing with publicity, banking, agricultural improvements and personnel.

These national interests obliged him to delegate authority to Thomas Gilbert, MP, his General Land Agent and personal assistant, or ADC, as he liked to call his successor, and such was the accord between the two that he would often not bother to dictate letters to him but simply give instructions which Thomas would issue in letter form.[1] But Thomas was also a very busy man – he was General Agent to Lord Gower, whose resident land agent was William Bill,[2] whose daughter had married his brother, John Gilbert, proving just what a closely knit family affair the canal idea was involved in. Thomas was busy with his innumerable reforms, he had his own estate to run at

Cotton, he was practising as a barrister in and out of Parliament and was involved in business and waterway projects too, though he also made it a part of his duty to act as chief auditor: 'He came to Worsley every Xmas 'till 1795 to examine and state His Grace's accounts, staying about 10 days and taking the General Accounts with him for his Grace's eye; he was often in haste as he was in Parliament . . .' wrote Robert Lansdale.[3]

So the burden of daily administration at Worsley was delegated to John Gilbert, on the next step of the managerial pyramid. He combined the dual roles of land agent and resident engineer – a more modern term, this, but one which aptly fits; true, many later resident engineers were considered less important than the consulting ones, but that stage was yet to come.[4] The Gilbert correspondence clearly shows[5] that, although Francis did sometimes write or dictate letters to John, he would seldom presume to write back direct to his employer except in cases of some urgency – so normally he addressed his letters to Thomas, who, as General Agent, would make the decisions himself if he thought proper, or consult Francis if he decided that the issue warranted it. But John was a busy man too. He farmed his own demesne at Worsley, and in 1760 he bought, with his usual foresight, the Golden Hill estate on what would later be the route of the Trent and Mersey Canal, in partnership with his brother Thomas, James and John Brindley, and Hugh Henshall, the brother of James Brindley's future wife.[6] John Gilbert owned and directed a blacklead mine at Borrowdale in the Lake District, and a factory for it at Worsley in partnership with his brother; he owned 'salt works at Marston, near Northwich, Limestone quarries at Caldonlow and Astbury; a colliery at Mear Heath, along with Mr Henshall of Longport, Limekilns at Hemheath and Ecton, [a] Colliery at Clough Hall',[7] as well as silver and copper mines near Alton, Potterdale, Keswick and Stanhope, carried on by his firm of Earl Carlisle & Co., in which the duke held shares, the earl after whom it was named being Francis's nephew and heir. About 1765 the duke also helped to buy John the estate at Clough Hall, immediately above Harecastle tunnel, as a reward for his services, for he treated his staff as generously as his workmen, and there Gilbert built himself a comfortable country house with some twenty bedrooms and the luxury of mahogany in all the downstairs rooms.[8]

An engineer and company director on this scale who was also farming and carrying out a personal industrial revolution in the Worsley soughs could not possibly direct every detail, so he too had his assistants – men like Robert Lansdale who earned about a hundred pounds a year and wrote, 'I have been with him *in* all the Mines and assisted at the General Paydays which were held in the Summer.'[9] John also delegated, to men like William Brough, Mathias Shelvoke and Thomas Kent, who submitted their accounts to him, but he sometimes called in consulting engineers like John Smeaton and James Brindley, whose expert knowledge in engineering and water control would help to solve specific problems. Brindley's knowledge of surveying was also invaluable, and though he was not on the payroll of either estate, he could be certain of regular jobs for the Gowers or the Egertons. Sometimes two engineers would be consulted – Smeaton with Brindley for calculating the sough flow for Parliament: much beyond Brindley's mathematical capacity, and probably beyond John Gilbert's too – but these were not regular employees, and when Brindley asked to be put on the Worsley payroll permanently the offer was not accepted, though he would happily have chosen the security of a fixed salary.

Brindley also was a very busy man, with his navigable soughs running from the Harecastle tunnel, planned on the lines of the duke's at Worsley, and often away on mill work or surveying, so he too had to delegate to foremen and relatives like Bennett and Harrison, and later to partners like Hugh Henshall who helped him as much with their cartography as he supported them with his garnered wisdom and enthusiasm. Such was the pyramidal structure of management for the Bridgewater canal, farming and mining enterprise – closer in many ways, ethical as well as administrative, to a medieval Honour rather than a modern corporation.

It follows from this that John Gilbert was resident engineer in charge of the waterway, and here he is supervising Brindley's work and earning some hard grumbles for his pains: 'he is davred to imploye ye carpenters at Cornhill in making door and window frames for a building in Caslefield and shades for the mynors in Dito and other things. I want them to Saill moor. He took upon him diriction of ye back drains and Likewise such lands as be twixt the 2 house and ceep [keep] [on the] uper side the large farme, and was displeased with such raing [range, or line] as I had pointed out.'[10] However furious Brindley may have been with these alterations to his survey line, a consultant engineer might argue with, but certainly could not override, a resident engineer in an organisation as highly disciplined and closely structured as the Bridgewater estate. Naturally, when visitors like Sir Joseph Banks were shown over the canal by John Gilbert, the resident engineer would be generous in his praise of his colleague, leading Banks to write, in a rather vague way, about 1766, 'These and many other usefull and ingenious inventions were thought of and invented by Mr Brindley who also did most of the Engineering work on the Canal he is a man of no education but of Extremely strong natural Parts he was recommended to the duke by Mr [John] Gilbert who found him in Staffordshire where he was only famous for being the best Mill wright in the Country.'[11] But Banks had a further statement to make relating to John Gilbert's supervision of the engineering of the canal.

As the cut progressed towards the Castlefield terminus on the fringe of Manchester the friction between the two engineers increased. Brindley, though not always an easy man to work with, was quick to make up quarrels on all but one point – his old horse. Being still a lonely bachelor, his naturally generous nature lavished all its affection on this mare, his constant companion on so many of the long journeys in which he had proved his skill as a craftsman and inventor. One evening when Brindley was in bed at Stretford John Gilbert's stallion – possibly the very one exchanged with the highwayman – broke into the field and got her with foal. For Brindley his means of transport was his livelihood, and he was mad with anger, entering in his notebook 'a meshender [messenger] from Mr G. I retorned the anser No more sosiety,' swearing that Gilbert, who did indeed have a sense of humour, had done it as a practical joke. The wrangle reached such a pitch of intensity that it interfered with work and threatened their Golden Hill mining partnership at Harecastle, so Thomas Gilbert and his eldest son Tom had to ride over to Gorshill to take the tetchy old engineer out on a drinking spree. Though placated, Brindley ever afterwards maintained that John had tried to prevent him from doing his work properly.[12]

But similar problems of temperament attend most important undertakings, and by 11 November 1763 Brindley was ordering the pit for the spindle of the mortar mill at Worsley, as fine a piece of work as ever he did and an invaluable contribution and also noting that Gilbert had ordered one of the twenty-ton boats to be available for his use the following morning.[13] As the canal moved forward a fleet of boats crammed with the gear of blacksmiths, carpenters, masons and toolmakers was drawn towards the Mersey. Earth and stone dug and blasted from other sectors were brought to the western front in ballast boats which dumped the refuse on the canal to build embankments across low-lying land, since the duke had stipulated that he preferred to pay the cost of keeping a level rather than fill his cut with delaying locks. Progress was slow because money was short, and when a board was nailed to the trunk of a poplar at Dunham Banks to show the route which the canal would follow the locals nicknamed the spot 'The Duke's Folly', believing that if ever the cut got that far it would find its owner a pauper.[14]

The mythical vision disseminated first by Phillips and then by Smiles, of Brindley as a dedicated genius solely responsible for driving through the work on the Bridgewater canal, is based more on the impressions of gongoozelers and visiting journalists than on the hard documentary evidence which every historian should consider first. In 1762, at one of the most critical stages of construction, he was away surveying Offerdene for Earl Gower,[15] or, a little later, the Stockport Canal for the duke. The account books show him working increasingly in a consultant capacity after 1762, while John Gilbert's salary was raised to £300 a year from 4 September 1762.[16] Despite a still popular belief that the duke's canal was constructed in one year[17] there is no possible sense in which the Bridgewater could be said to have taken less than seventeen years to build, but if we allow for the Liverpool docks it can be argued that forty years is not too high an estimate. Since Brindley died in 1772, and was constantly busy with such projects as the Trent and Mersey from before 1765, it is clear that his part in this immense enterprise was distinctly limited, though doubtless important. Certainly he captured the public fancy, and used publicity extensively for furthering the work of his engineering partnership.

Like a silver snake the canal moved forward from Worsley, winding towards Manchester, crossing Barton aqueduct in July 1761, vanishing underground at the Delph, and only checking on its westward course over Chat Moss, since the second and third Acts had substituted for that route to Hollin Ferry, down river from Irlam, the southern one across Cheshire. The Chat Moss line had been intended to carry the canal to the Egertons' traditional markets in the Hollin Ferry area, and provide additional water.[18] By September 1763 the duke's men were unloading and his agents selling his first cargoes of coal at Cornbrook,[19] where the Gut linked his canal to the navigation, but it was 1764 before he managed to settle the Hulme Hall estate of George Lloyd and struggle through to Castlefield.[20] Worsley coal did not reach Preston Brook until 1771, and the Runcorn locks were not completed until Lady Day 1776, when Brindley had been dead four years. Meanwhile the duke continued to sell coal in Salford by lowering it on the gig crane to his growing fleet of flats on the old navigation, a practice which continued throughout the century.[21]

The canal was entering ever more hostile territory as it crept into the environs of Manchester. The duke had been obliged to buy Lloyd's house because one of his Acts forbade him from cutting within thirty yards of Hulme Hall, nor was he allowed to touch more than a corner of the territory of Edward Byrom, chairman of the Old Navigation; land prices ran very high indeed at a time when his 'finances were strained to the uttermost'.[22] The Castlefield docks and warehouses rank among the most brilliant engineering achievements of the triumvirate: Brindley harnessed the waters of the Medlock, a tributary of the Irwell, with a circular weir which coped with much of the surplus water, though Arthur Young in 1774 criticised it as being subject to flooding.[23] Young noted the starvationers, 47 ft long and of 4½ ft beam, each carrying a dozen metal coal boxes with lifting handles, and holding 8 cwt of coal. A sough had been delved into

Castlefield hill to draw up these coal buckets, which saved the carters the long haul across the western slope. Banks described how Brindley, with his knowledge of mill work, had constructed a crane to draw containers up a shaft from the sough 'on the same principles as mill tackles . . . called the Endless Rope', and driven by water from the Medlock. By throwing the upper wheel in or out of gear the endless rope could be tightened or slackened to drag up the heavy coal containers.[24]

It was reckoned that two men and a boy could shift about five tons of coal in rather less than an hour by this means, and Woolley, the duke's Manchester agent, would sell it to the public, generally hungry for fuel, always giving precedence to the poorer people. The triumvirate had scored another notable invention with almost certainly the first example of containerisation – Castlefield was regarded by contemporaries with much the same wonder as Barton or the soughs.

Two vast warehouses were built about 1765 – the duke's, later renamed the Bridgewater, or Old; and Henshall Gilbert & Co's (later the Grocers'), for their firm carried substantially as the official boatmen of the Trent & Mersey Canal Co., and needed much storage space. There appears to be no record of Worsley coal sales at Castlefield before 1 August 1765, when the first wharf came into use, so it seems that most of it was being sold up the Old Navigation to Salford on flats loaded by the gig crane over Barton, until the Castlefield harbour was ready. These flats were repaired by being hauled up the Gut at Cornbrook and towed to Mugg's shipyard below the Delph, or to a growing repair yard at the inland port of Castlefield.

It is hardly surprising that the duke should have met with opposition as his waterway entered the territory of those Manchester men whose traditional loyalties lay with the Mersey & Irwell company. It would seem likely that they had, in their despair, offered their navigation to the duke for about £13,000,[25] but this had been rejected, since Francis was not only said to have opposed monopoly in principle, but had also used powerful arguments against monopolistic practices in obtaining his parliamentary Acts. The duke was willing to join forces with the Salford Quay carrying company, but this was understandable, as they were both important hauliers on the Irwell route. The other major carrier was the 'Old Quay' or the Mersey & Irwell company, carrying in its own name. Within a few years he had bought a substantial number of Salford Quay shares, and became the sole owner in 1779. He did hold a few shares in the Mersey & Irwell – just enough to provide information.[26]

By 1763 Brindley had almost completed Stretford aqueduct across the upper reaches of the river Mersey, and the cut was heading south-west for Altrincham. He generally lunched at the Bull Inn at Stretford for 8*d* but he was not the only consulting engineer employed by the duke and Gilbert, for one correspondent chides another for giving Brindley all the credit, pointing out that 'The Duke of B. has another ingenious Man, Thomas Morris, who has improved on Mr Brindley and is now raising a valley to the Level by seven double Water-Locks, which enable him to carry earth and stones as if down steps. When each Lock is opened it admits a loaded vessel on one side, and lets out an empty one on the other; by which means Tons of Earth are carried, and the Valley will soon rise to equal the hills around, and the navigation keeps its level.'[27] He added that within three months the cut would reach Lymm and that the Earl of Stamford had built himself a 'bathing-house' on the canal bank to utilise to the full this valuable new amenity which his friend and fellow peer had provided for his recreation.[28]

Joseph Banks's evidence

This particular letter is tolerably accurate, but far too much attention has been paid to the statements of casual visitors and journalists in no way as reliable as the careful scientific observations of Sir Joseph Banks (1743–1820), the brilliant botanist who accompanied Cook on the *Endeavour* in 1768 and was elected President of the Royal Society ten years later. The year before he sailed for the Pacific he travelled through Wales, the Midlands and the North West, searching for rare plants and herbs, but noting many other things with the trained eye of the scientist.[29] After spending the night at Knutsford he sallied forth to see the earth-raising boats at work on the cut. He noted two different types, 'Those called Fly boats which are most used are constructed by Joining two Boats together at the distance of about three feet, which is done by cross beams of timber.' The earth was piled high and walled in between the two hulls, while small doors held by chains could be released to slip this soil down to the bed of the canal. The others, called diving boats, worked on a similar principle but were only single hulls divided into partitions with doors in them, half being buoyancy tanks.[30]

He rode on to Worsley, visited the mortar mill and noted its workings with care, making minute sketches of the mechanism, adding that the duke's mortar 'certainly is stronger and sets quicker either under (or over) water than any other in the land'.[31] Cement was not then used, and lime had to be carefully ground and blended with sand for the miles of tunnels, hundreds of bridges, and dozens of offices, wharves and new buildings that Gilbert and the duke were hard at work on. He delved into the heart of the 6,000 acres of Chat Moss, searching for rare plants. 'The duke has been six or seven years attempting to drain the parts of this moss which belong to him, with no great success at present, tho' he has all the hopes possible of Compleating it in time' (i.e. given time),[32] but the peat was so soft that drains cut in it ran together almost at once, though the waste from the mines was gradually building firmer areas. The following day he spent 'in viewing a discovery of Lime made by Mr Gilbert when first he came here – a very valuable one to the Countrey, as they were used to bring Lime miles, except the Sutton, which was sold too dear for the Farmers use'.[33] The Sutton lime came, he stated, from a quarry at Astley, near Worsley.

It was winter, and the canal was frozen over, about an inch thick, but Gilbert had his ice-breakers out – 'a broad stemmd boat in which were 7 or 8 people who swayd her with great force as she was drawn on by the mule & at the same time struck any Large peices of ice in peices with clubs they held in their hands'.[34] He recalled that one of the most telling arguments against the duke's canal was that it would be useless in winter when even moderately iced over. This was dramatically disproved during Banks's visit by the ice forming hard enough to carry a man during a bitter frost; the duke's flats sailed triumphantly through the passages cut by his breakers, but nine barges on the old navigation 'gave out a little above Barton Bridge, attempting with all their Powers of men and horses to get up, but in vain. Whether made more Eager to accomplish their Endeavours by seeing the canal navigation going on without interruption, I cannot tell, but one of them was drawn against the ice 'till she had a hole cut in her bottom which would have sunk her, had she not been immediately unladen.'[35] Far from casually visiting the soughs, he spent seven exhausting hours studying the stall-and-pillar system of mining in the greatest detail, noting particularly how the majority of the pillars supporting the roof were removed as the men worked on through the 'rank', the ceiling being allowed to cave in. At the time of his visit, the winter of 1767–68, he noted that the main sough level was almost a mile long, with one cross-level of about 1,000 yards, wide enough for two boats to pass each other.[36] Of the canal above ground he wrote, 'the whole intended Length . . . from Worsley . . . is 34 miles of which 21 are Compleated, & 13 Remain unfinished', and it carried vessels of forty tons and upwards, while he praised the canal idea for providing a safe carriage for goods in time of war, when privateers could attack coastal shipping.

John Gilbert had spared Banks some of his limited time, and it is to this that we are indebted for a statement which finally resolves the vexed question of the exact roles played by each member of the triumvirate in the construction of the Bridgewater canal: 'we must acnowledge ourselves indebted to its noble author, & not a little to his cheif executor, Mr Jno Gilbert, whose most indefatigable industry, himself overlooking every part, and trusting scarce the smallest thing to be done except under his own Eye, I myself have been witness of'.[37] Later he pays the tribute already mentioned to Brindley for his work on the winding gear at Castlefield, and for other unspecified inventions, as well as much of the engineering on the earlier stages of the open-air cut. From this it is abundantly clear that Francis, the 'author', was personally responsible for the canal which bore his name, that John Gilbert was the engineer in charge of all operations, and that Brindley also played a part for a time.

Banks also visited Brindley's Stretford aqueduct over the upper reaches of the Mersey, noting the care with which the men were, in that intense frost, washing the stones in boiling water before they laid them, to repair the arch which the savage floods had washed away. Earlier he had taken a look at the work on Brindley's famous Harecastle tunnel, noting that it had been driven 'about 100 yards, is well arched & nobly sized, but their mortar is so soft & seems to have so little care taken in making it, that I cannot help having my fears of accidents that may befall it when it comes to bear a large weight of hill'.[38] This, combined with other evidence from the canals which Brindley supervised, substantiates the view of the grandfather of the sage of Downe, Dr Erasmus Darwin, that James Brindley 'was better qualified to be the contriver rather than the manager of a great design'. Nor should the duke's part in his own waterway be underrated. Arthur Young, a shrewd observer and no mere sycophant, described Francis as 'so bold and daring a genius. To see him engaged in undertakings that give employment and bread to thousands, that tend so greatly to the advancement of agriculture, manufactures and commerce, of an extensive neighbourhood . . . must command our admiration; his Grace has a mind superior to common prejudice . . . one of the truly great men, who have the soul to execute what they have the genius to plan.'[39]

The duke's lieutenants also needed to be versatile men. Gilbert, Morris, Brindley and Kent, who was in charge of the coal mines, had to be skilled land surveyors, administrators, mining engineers, carpenters, masons, brickmakers and paymasters, as well as accountants. By 1765 a visitor to the western front found 400 men putting finishing touches to the cut near Stretford, and work had already begun on the Cheshire bank of the Mersey at the projected terminus. Brindley, though busy elsewhere, was sometimes called in for advice and guidance, especially in matters of water control, but Gilbert soon had the help of his son John, who had also received an excellent training at the Snow Hill and Soho works in Birmingham, under the aegis of his father's old friend, Matthew Boulton junior.[40] Nor was this an entirely one-sided exchange, for in 1765 Boulton, ever alive to any new project, was 'in the Whitsun week seeing the Duke of

Bridgewater's works near Manchester' in the pleasant company of his old friends, and doubtless reporting on young John's progress as an apprentice manager.[41] Within a year or two Boulton was introducing the canal idea to Birmingham.

A concrete example of the elder John Gilbert's work as resident engineer in charge is the dramatic letter to his brother Thomas when Brindley's Stretford aqueduct collapsed. It was penned 'On Board the Counting House on my way to Stretford', this being the pay barge for the navvies.[42] A 'flu epidemic was raging and killing off thousands, and most of the duke's men were ill, but time was precious, and the horses were sick too – 'we are obliged to work them, and think moderate work does them good'. Banks had commented on this Stretford flood as far the worst within living memory, and Gilbert described how it had come 'with great violence from High ground . . . with such force, when it checked at the entrance of the Arch it broke up the foundation weirs 8 foot below where it was set from', several piers collapsing, though one held firm; 'but this', he adds, 'we must talk over with Mr Brindley when he comes'.[43] Brindley cannot be in any way blamed for a natural hazard of this magnitude, but Banks's description of Gilbert heating cauldrons and working pumps to repair this vastly expensive and time-wasting damage in the heavy frost is an engineering saga as vivid and spectacular as any recorded. John Gilbert was often at his best when facing the challenge of a really severe emergency.

Notes to this chapter are on p. 190

10

The Trent and Mersey

'Difficulties are the boundaries of narrow hearts,' wrote that veteran waterway pioneer Francis Mathew, in 1655.[1] The duke was not the man to run from a sea of troubles, and he and Gilbert fought their dearth of working capital with every device they could think of. On the side of the cut across Trafford Moss they planted trees which sprouted quickly and were soon being cut as drying poles for the Manchester dyers – a small economy, but it strengthened the banks and brought in a little more revenue.

For the duke himself, hard work and hard riding had become a substitute for the savage drinking of earlier years, though he kept his cellars well stocked and continued to entertain quite generously. A heavy smoker of long churchwarden pipes, he was restless and irritatingly energetic, filling any lull in the conversation by jumping up to tap the barometer in the hall. It was not just commercial but also political business which obliged him to travel widely, and on his way through Salisbury, probably on a visit to the Thynne family at Longleat, he passed the evening watching a performance by a group of strolling players when, to his surprise, he realised that one of the actresses was the pretty Miss Langley, daughter of a tenant farmer on one of his estates. The girl had acquired a romantic passion for the theatre and had run away from home with the leading man, only to discover that show business was a sterner matter than she had imagined. As the takings had been meagre, she was extremely hungry and deeply grateful for the ample meal which was promptly provided for her at the duke's inn. Her parents had died during her absence, so Francis set her up on one of his farms and gave her an allowance, though there is no certain proof that the scandal sheets of the time were correct in assuming that she was his mistress; if they were, he had at least learned more about discretion since his youthful episode in France than many of his fellow peers evinced in that particular age. His kinsmen and friends were fond of him but quite unable to grasp the vital importance of his work: 'The family generally addressed him (Lady Georgina, I think, always) as Dux, to which he good humouredly, & as it were habitually responded. Mr Sparrow passed some days with us, and during his stay, the Duke was every evening planted with him on a distant sofa, in earnest conversation with him about canals, to the amusement of some of the party.'[2] Eating well was not a passion with him, but he enjoyed good food and grew fatter. As a gambling man he appreciated that the odds against his sudden death were gradually lengthening, and his hard-riding, open-air life was a recipe for continuing good health.

As his canal wandered very slowly westwards it became clear to those with sufficient insight that the technical problems of the canal idea, once thought unsuitable for cold climates, had been substantially overcome. They also began to have some inkling of the social benefits likely to accrue from nearly halving coal prices, supplying and carrying for industry and, closest of all to the hearts of squires and farmers, providing far better communications for agricultural supplies and produce. This was all very altruistic, but there was still some doubt as to whether there was money to be made out of canals, and the duke was doing far from well on that score.

The idea of linking the main ports of England – Bristol, Hull, Liverpool and London – with a canal network had been a subject of speculation for centuries in various forms. In 1655 Francis Mathew had urged Cromwell to link the Thames to the Severn 'for the Advancement of Trade and Traffique', personally exploring the remotest water byways to corners where 'the brook was scarce as broad as my Boate', and struggling past the innumerable mills which blocked the rivers of his day.[3] By the eighteenth-century every Englishman who pretended to any proper education knew of the

French achievement in linking the ocean ports by inland waterways, while in 1755 'The late Mr Hardman, an intelligent merchant of Liverpool', was the chief promoter of a survey of the Trent and Mersey line by Taylor and Eyes which preceded any others.[4] Then the duke and Earl Gower took over and planned the route of the Trent and Mersey along the lines that it would later run: this is confirmed by Fenton's statement that

> The following map is definitely in the hand writing of Mr Hugh Henshall – a map of the Grand Trunk canal, very nearly as it now exists, as extends from the Trent at Wilden Ferry in Derbyshire to Longport – with three branches viz. the one afterwards made by the Coventry Coy. from Fradley Heath . . . The Plan is on Vellum & contains References with the names of landowners and the lengths through each property. Only the Title is not in Hugh Henshall's writing & reads: 'A Plan for a Navigation chiefly by Canal from Longbridge near Burslem in the County of Stafford to Newcastle, Lichfield, & Tamworth & to Wilden in the County of Derby.
> By James Brindley
> Revis'd and approv'd by John Smeaton 1760'
>
> prob. this is written by Smeaton.
>
> The Act for the Grand Trunk Canal did not pass 'till 1766, but you see that the Plan for the Staffordshire part of it was complete nearly as soon as the Duke obtained his first Act.[5]

From this it is clear that the Egerton–Gower interests had the Trent and Mersey neatly planned far in advance of Wedgwood taking up cudgels on its behalf, and were simply waiting for time and public opinion to ripen. Brindley had been employed for his skill in seeing the lie of the land which a cut could follow, Henshall had, as usual, drawn his map for him, and John Smeaton had, as with measuring the sough water, supervised his colleagues and given their work the imprimatur of a professionally skilled engineer. There is at present not the slightest shred of evidence to support the theory that Brindley, a millwright and occasional consulting technician to these estates – a man of limited education – would have been able to originate the Trent and Mersey concept, which had been evolving in the minds of Liverpool merchants for at least six years. The canal idea ranked high in the priorities of politicians like Earl Gower because the coasting trade was vulnerable to pirates in peace and privateers in wartime. George, Lord Anson (1697–1762), who had shared the expense of Brindley's initial survey in February 1759,[6] was a tough old seadog who had circumnavigated the globe, carried out a brilliant raid on the Spanish colonies and soundly defeated a French fleet in 1747; he was, for a while, Gower's neighbour in Staffordshire and would certainly appear to be a more likely candidate for the honour than Brindley.

Charles Roe

There were local as well as national politics involved in the Trent and Mersey scheme for a canal to link Liverpool to Hull. Josiah Wedgwood, and the men who were recreating the pottery industry with work of exquisite skill and genius, were determined that the cut should pass their doors, so that a scheme to run from the river Weaver by Nantwich and Stafford to the Trent at Wilden Ferry – now Cavendish Bridge near Shardlow – met with their fiercest opposition: it would never have catered for their interests. Another plan, supported by Charles Roe of Macclesfield, advocated an elliptical cut from the Weaver, through the town of Stockport to Manchester, and this caused some furrowed brows indeed among the duke's party. Roe's proposals would have cut the duke's southern flank, drawn off the vital trade to Manchester and Salford, and left him to recoup this loss as best he could from the Liverpool line, since the crippling limitation on coal rates made his mines a very long-term asset. The duke's only possible solution was to go in with the new proposals, join forces with these innovators and ensure that if they did build their cut it should link with his own. Thus it came about that the duke allied himself with the Stockport squires and traders, and presented a petition stating that he had taken a survey – Brindley had been chosen for this and was generously paid for it by the duke on 28 October 1766[7] – from Sale Moor to Stockport; all this would, it was claimed, serve many Yorkshire and Derbyshire towns, and – a sure sign of the duke's personal hand in the matter – increase the 'general & public Utility . . .'.[8] The Stanleys may have been involved, as hereditary stewards of Macclesfield and opponents of the duke.

Edward Byrom, Kenyon and other Manchester merchants holding shares in the old navigation were horrified, and wrote urgently on 9 February 1765 to their old friend and parliamentary ally Lord Strange:

Your very earnest endeavours to serve us in opposition to his G. of B.'s last Bill for extending his Canal will always be gratefully remembered tho' that opposition was ineffectual. We now are more alarmed with an intended application to P. by the principal Inhabitants, Traders &c of Stockport . . . to communicate with the D. of B.'s canal – the Utility of this to the Public we do not presume to determine but what will consequently materially affect us is that the whole river Mercy may, according to the Plans which have appeared, be taken into the intended canal – In which case our Navigation will be rendered useless in the dry seasons of the year – we have applied by our Agent to Mr Newton of Stockport who is to sollicit this Bill to desire he would give us a Plan of the intended works – In the meantime we have adressed ourselves to you hoping for your Assistance in case an attempt be made to rob us of the supply of water from the Mercy – We are with the greatest regard – Sir, your Most obliged humble servants –.[9]

Almost a year later the opposition had shown its hand, and on 21 January 1766 a London solicitor wrote to Allen Vigor, an attorney at Manchester, stating that, on the same day that the duke's and the Stockport petitions went to Parliament, 'was another Petition presented from the Gentlemen of Manchester, Stockport, Macclesfield, Knutsford & Northwich, for a Navigable Canal from Wilton Bridge to Manchester by the way of Knutsford &c., which we apprehend from your letter must clash with the Interest of Messrs Byrom, Kenyon &c.'[10] Certainly this waterway would have met with the combined opposition of the Gower–Egerton interests also, but the duke was probably keen on the other Stockport link, and may well have begun cutting it, the line having been surveyed by Brindley by 8 January 1762.[11]

The trustees of the Weaver Navigation were determined that both the Trent and Mersey and the projected Macclesfield Canal should lock down into their line and enrich it with an abundance of tolls, from the prospering port of Liverpool, with its sugar, tobacco and cotton imports, fed by Sankey coal to the potteries and salt towns, thus forming a new triangular trade, which contributed as much to the port's prosperity as the more sinister slaving triangle which ran south and across the Atlantic.[12] But the duke's supporters knew that he was short of capital – they realised that it was vital to his interests that he should link with the Trent and Mersey, and that was, in any case, the plan which Francis had evolved long before Charles Roe and the Macclesfield merchants managed to get their Act through the Commons, but failed in the Upper House.[13]

Josiah Wedgwood

Meanwhile the first priority was for the various contestants to pick their allies and become better acquainted with each other, so on 6 July 1765 Josiah Wedgwood and his friend Sparrow, an able solicitor and a keen canal man, called on the duke at Worsley, taking their plans with them. 'We were most graciously received and spent about 8 hours in his G——'s company,' wrote Josiah, '& had all the assurances of his concurrence with our designs that we could wish.'[14] Francis showed the master potter an urn which had been uncovered by the excavations at Castlefield, and personally conducted them around his busy docks and soughs. When he cared to take the trouble, and canals were at stake, none could be more charming or hospitable than their host, and before they left he ordered a complete table service of china from Wedgwood which served as a prototype for the more famous one turned for Catherine the Great of Russia, and rediscovered in the Hermitage at St Petersburg in 1909. The Russians expressed some surprise that any of this should have survived, as Catherine was addicted to throwing china at her servants.[15] A painting on one of these pieces, of the canal at Worsley showing the duke's gondola, and another of the aqueduct, was taken from sketches by the French philosopher, Jean-Jacques Rousseau. By the kindness of the Davenport family he had been lent Wootton Hall in Staffordshire, where he stayed with his mistress, le Vasseur, and wrote his curious 'Confessions.' When Francis bade adieu to his new allies he ensured that they had what Josiah described as 'the honour and pleasure too of sailing in his Gondola nine miles along his canal, thro' a most delightful vale to Manchester'.[16] Francis was a keen promoter of pleasure boating, and encouraged his friends to build yachts, which he always preferred to call gondolas, in honour, perhaps, of his guardian's years in Venice – a growing number were moored at Castlefield as the cut progressed.

The Gower–Egerton–Wedgwood lobby, known unkindly to their opponents as the 'schemers of Staffordshire',[17] powerfully opposed the Macclesfield canal Bill, but though it passed the lower House, the duke was there to meet it in person in the Lords, aided by Brindley's technical evidence, and was present at every meeting at which the Bill came up. He even offered the Macclesfield promoters a branch line from his proposed Stockport canal, but it was well known that he was reducing work on his own cut for lack of funds, and the offer was firmly rejected.[18] The Macclesfield coal mines were almost worked out, and cheap fuel by canal was needed at once, not in seven years' time. Charles Roe of Macclesfield (1715–81), who gave evidence for the scheme to outflank the Bridgewater, was described by his niece as 'much *feared* and much *loved*', but he was a man of considerable stature who had built up mining, smelting and other trading interests in Scotland, Wales and Ireland, and he and his supporters were no mean opponents, so that the duke's party was faced with a severe struggle. Brindley's technical evidence, admirably presented and backed up by the best counsel, showed clearly that the cost of carriage would be lower on the duke's lockless canal, and quicker, and that the Macclesfield, if built, would detract from the precious waters of the feeder streams supplying the Bridgewater. The accusation that the duke wished to obtain a 'stranglehold' on the Trent and Mersey when in fact he was undertaking, for a reasonable price, to save the proprietors substantial sums on construction, and the calumny of some biased letter writer that he wished to become the largest waterway carrier in Europe, are hardly acceptable now.[19] The Gower–Egerton plan for joining the main ports was a national one designed to benefit England, while the cut advocated by Charles Roe was designed chiefly to help Macclesfield, so that the decision of the Lords in April 1766 that the Bill should be postponed was not perhaps unjust.

Brindley had been the obvious choice as engineer for the Trent and

Mersey project, partly because his infectious enthusiasm would supply the inspiration required for such a vast enterprise, partly because his work under John Gilbert and the duke had taught him much, but chiefly because he was a close friend of the Wedgwoods, who were very anxious to cut expensive overheads by using the new canals to obtain cheaper coal for their expanding furnaces, china clay from the south coast, and flint from various sources.

It was therefore clear, in Wedgwood's phrase to his partner Bentley in Liverpool, that 'The Duke (from what I have heard him say) would much rather come to you all the way by a canal than down the River',[20] the river in question being the Weaver – meaning that Francis, no lover of navigations, was bent on linking his canal arterially across the country with the Trent and Mersey. Bentley was pressing for an aqueduct high over the mouth of the Mersey to carry the new cut direct to his city of Liverpool without risking the tides, storms and shifting sandbanks of the estuary, but when Wedgwood and Brindley met the duke and John Gilbert at Trentham, Francis and his agent, as Wedgwood put it, 'looked at each other as though some secret design of their own had been discovered by another before the time they thought proper to avow it'.[21] Shrewd though Wedgwood was, neither Gilbert nor the duke would ever have toyed with such an impracticable idea as a vastly expensive aqueduct over the lower Mersey, knowing as they did the weakness of Brindley as a theoretical engineer, and the flagrant inaccuracy of some of his predictions when Smeaton was not available to aid him. It must also be remembered that Francis was so desperately short of working capital that he had had to raise the mortgage of £25,000 from Child's bank in London, and was still uncertain how far he was in a position to bid against the Weaver–Macclesfield opposition, so he sent John Gilbert to Burslem 'to give us some further hints, instructions and encouragement in our design, and amongst other things desir'd your scheme (the aqueduct plan) might not be mentioned at present, and with a significant nod told me great things might be done at a proper time, but we had enough on our hands at present'.[22] Meanwhile Bentley's publicity machine was at work on a pamphlet penned by Dr Erasmus Darwin. Gower pompously expressed his horror at the impropriety of a reference in it to gondolas gliding along the canal as an added attraction – a phrase slipped in by the duke – and, as Strachan Holme pointed out, the only light relief in an otherwise tediously technical political broadside.

Aqueducts and barrages

Inspired by Bentley's idea, Brindley had dreamed up ambitious plans for a combined road bridge and aqueduct crossing the Mersey upstream of Warrington, but the cost would have been prohibitive when combined with negotiating with the hard-bargaining proprietors of the Sankey Navigation for crossing their cut as well. Moreover the beam of the 'dukers', as the Bridgewater flats were later termed, would have excluded them from running on Brindley's narrow canals.[23] The aqueduct idea was therefore almost as impracticable as George Merchant's brainwave in a letter addressed to the 'Mare of Liverpool' on 17 December 1768, suggesting nothing less than a Mersey barrage 'above the new dock to the opposite shore of sufficient strength to resist the tide', containing 'gates' for shipping, and adding, 'please send Master Bridley [*sic*] a coppy of this & desire his opinion – from his caracter I have no doubt of his acquiescance'[24] – an assessment which history has now, in some measure, confirmed.

At last all was ready, and on 21 December 1765 Brindley was able to write in his progress report to John Gilbert at Worsley, that Sparrow and his allies were gaining landowners' consents from Harecastle to Agden (his spelling has been partly modernised):

On Tuesday Sir George [Warren] sent Nuton into Manchester to make what interest he could for Sir George, and to gather ye old Navigators to meet Sir George at Stoperd to make a Head against His Grace.

I saw Doctor Seswige who says he wants to see you about payment of his land in Cheshire

On Wednesday there was not much transpired, but was so dark I could scarce do anything

On Thursday Wedgwood of Burslem came to Dunham and sent for me and dined with Lord Grey & Sir Harry Mainwaring and others. Sir Harry could not keep his temper. Mr Wedgwood came to solicit Lord Grey in favour of the Staffordshire canal, and stayed at Mrs Latoune all night and I went with him, and on Friday set out to wait on Mr Egerton and solicit him. He says Sparrow and others are endeavouring to get ye landowners consents from Harecastle to Agden.

I have ordered Simcock to ye length falls of Sankey Navigation.

Ryle [Royle] wants to have coals sent faster to Altrincham, that he may have an opportunity to drain off ye Sale Moor canal in about a week's time.[25]

The Gower–Egerton plan for a waterway system running from Worsley and Manchester, via Trentham and the Potteries to Lilleshall, Coalbrookdale and Bristol, was now plain.

By 30 December 1765 the Gower–Egerton–Wedgwood lobby was able to convene a meeting at Wolseley Bridge, near Lichfield in Staffordshire, to gain further support from local squires for their line from Trent to Mersey. Lord Grey had been won over, as well as Anson, the MP for Lichfield, Ashton Curzon and Thomas Gilbert, among many others. Gower opened the proceedings with 'a very sensible and elegant speech', and Brindley explained that the result of his and Smeaton's surveys was a plan for a cut some ninety-three and a half miles long linking the Trent to the Mersey by a line running south of the mountainous Peak District. The high ridge at Harecastle would be pierced by a tunnel the better part of two miles long from which soughs on the Worsley model would radiate into the Golden Hill coal and mineral mines. Such a waterway would need some seventy-six locks and about twenty-three miles of branch lines to connect important towns like Leek and Derby to the main system.[26] Heading the list of shareholders were Gower and the duke, with many names later famous in the annals of the industrial revolution – Gilberts, Wedgwoods, Brindleys, and the younger Matthew Boulton included. Samuel Egerton contributed £42,750 of the total of £300,000 worth of shares – a proportion so vast that it almost made this canal a family undertaking.[27]

The Bill for the Trent and Mersey was submitted to the Commons in February 1766 and met with immediate opposition. The Old Navigators, packhorse interests and waggoners combined against it with the coasting trade and corporations of cities dependent on road transport; but the support was also powerful and varied, even including boating enthusiasts looking forward to what the duke, riding his hobby horse again, had described as 'The amusements of a gondola that may convey us to many flourishing towns, through the most delightful vallies in the Kingdom'. On 14 May 1766 the Trent and Mersey Act was passed, and by a coincidence the Staffordshire and Worcester, surveyed by John Gilbert with his son to aid him, and steered by him through both houses of Parliament,[28] was approved on the same day, forging the vital link with the Severn, which would join Bristol to the Liverpool–Hull axis. Gilbert's salary for this work was the sum which, as the duke's accounts put it, the proprietors of the Staffordshire 'Allowed on that occasion in their account with his Grace for his share of the Expences of that Act'.[29]

The stage was now set for linking the rivers and three major ports as described on the duke's map, [30] but Brindley was not an engineer of the same calibre as Gilbert, nor did he have the duke's administrative ability to support him. While the Bridgewater Canal was built broad-gauge for the dukers and ran on the 81 ft level, avoiding locks by using embankments and a few contours, Brindley's cuts were narrow-gauge and frequently followed the contours rather than braving the engineering problems of tunnels and cuttings. True, Brindley was no longer working directly for a single aristocratic proprietor but for shareholders hoping for some return on their outlay, and was therefore obliged to conserve capital; yet the narrow gauge was a tragic decision which the nation would soon rue, for it meant that the wide boats of the northern and southern waterways could not enter the narrow levels of the Midland canals, and had to unship their cargoes. Had Francis not been so busy with the Macclesfield and Brooke opposition, and so hard pressed financially, he might have found time to oppose a decision which could not have been in the national interest. Though water was short on the new line it was shorter still on the duke's cut, which drew its main supplies from the Medlock at Castlefield and from the soughs at Worsley, passing, like the Sankey, above most, though not all, of the feeders; but Brindley and his partners, obsessed by the canal idea, took on more than they could manage or his declining health contend with.

Meanwhile the duke was still in dire financial straits. By 1765 the gross profit from his coal mines amounted to £5,230, but wages and materials came to £2,468. Subsidiary profits from lime, passengers on the sough and minor sales brought in another £1,000 a year, but his debt for the canal had reached the startling figure of £60,879, and was growing daily.[31] Even the £25,000 mortgage from Child's bank was soon submerged in the voracious waters of his canal.[32] Under these circumstances his offer to bring the line of the Bridgewater south to join the Trent and Mersey at Preston Brook was not ungenerous, adding as it did permanent costs and mileage to the Bridgewater.

When the news spread that the Trent and Mersey Act had passed there was great rejoicing in the Potteries. At a solemn ceremony in July Josiah Wedgwood cut the first sod and local landowners vied with each other to wheel a barrow or turn a lump of turf. Brindley, with his usual sublime self-confidence, assured the proprietors in 1767 that 'The whole of this Navigation will be finished in five years', scorning the puzzled protests of a few Jeremiahs who timidly wondered how ninety-three miles and more could be covered so soon. Had he lived until the opening of the Trent and Mersey in May 1777 he would have lost his £200 wager by double the margin. Yet his closing years were brightened by his marriage in December 1765, at the age of forty-nine, to a girl of eighteen – Anne, daughter of John Henshall, the land surveyor, and sister of Hugh, who drew the old engineer's maps for him, and became his partner on the many waterways which he surveyed.[33]

By July 1767 Brindley was hard at work cutting the 2,880 yard long Harecastle tunnel, roofing it, as we have seen, none too well, struggling with severe flooding, and driving soughs into the Golden Hill mines radiating from the main passage. The accusation that Brindley and the Gilberts used their influence dishonestly to carry the tunnel through their own mineral rights is completely unacceptable: the ore extracted added generous revenues to the tolls, and it was therefore rational to exploit it. The ill feeling more probably stemmed from the delicate ethical point of just how much time and labour was spent on the partners' adits and how much on the proprietors' tunnel. Yet Brindley's surveying skill had induced him to pick the very spillway through the hills likely to gain him the best levels for water supply.[34]

Fame came to Brindley very late in life, when he was already dying of overwork and diabetes. His forthright character, able but blunt plain-speaking and hard-drinking, earned him close friends and bitter enemies. He and his partners seem to have done little, if anything, to eradicate the growing myth that it was Brindley alone who had engineered the duke's cut, despite his having been almost fully occupied in carrying the arterial canal idea out in England from 1765 onwards.

Even Josiah Wedgwood, though one of Brindley's most fervent friends, and grateful to him for the inspiration of his eloquence, was not uncritical. 'I think Mr Brindley – the *Great*, the *fortunate*, *money-geting* [*sic*] Brindley, an object of Pity! . . . He may get a few thousands, but what does he give in exchange? His *Health*, and I fear

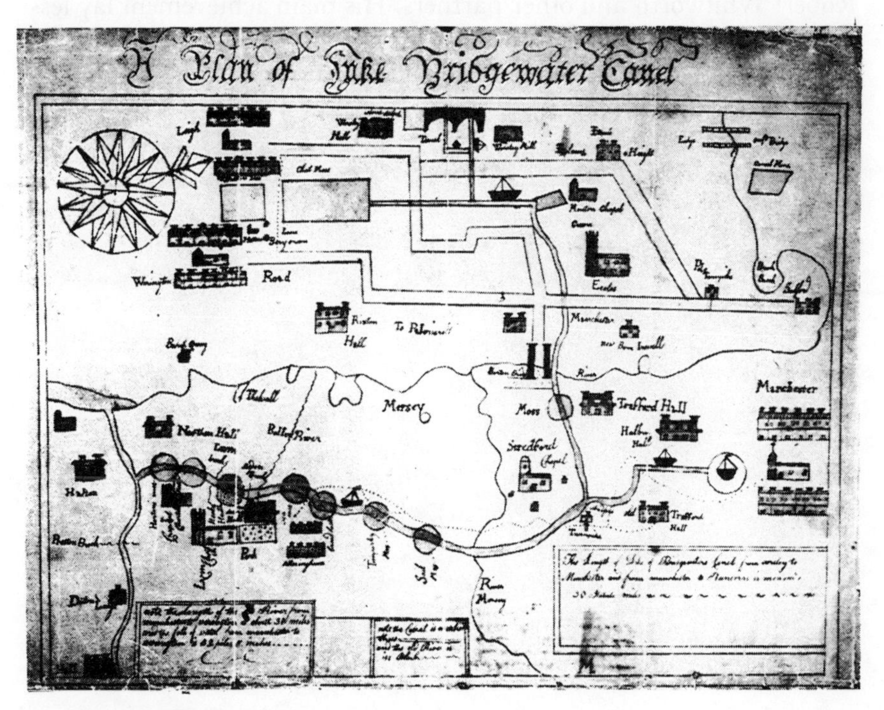

14 Brindley's only known map suggests that he was more nearly a craftsman than an engineer

his *Life* too.'[35] Dr Erasmus Darwin, who diagnosed his diabetes, though too late to comfort him, agreed with those who maintained that he was 'better qualified to be the contriver rather than the manager of a great design'.[36] With, it is said, some twenty-two waterway projects in hand to meet the canal boom, Brindley and his partners were gravely overstretched. His final illness was touchingly recorded by the indefatigable Wedgwood: 'I have been at Turnhurst almost every day this week, and can give you but a melancholy account from thence. Poor Mr Brindley has almost finished his course in this world. He says he must leave us and indeed I do not expect to find him alive in the morning.'[37] When he died, aged fifty-six, on 27 September 1772, he had completed only three canals,[38] leaving his life's work to be carried on by Hugh Henshall, Samuel Simcock, Robert Whitworth and other partners. His main achievement lay less in the mathematical precision of his engineering than in his dedication and devotion to that cause which gave England a more effective transport system.

Notes to this chapter are on p. 191

11

The battle of Norton Priory

While Brindley was puzzling over the problems of Harecastle tunnel and spurring his old mare from one waterway project to another, the duke had noticed that local people were using the barges increasingly to get to and from market, so he obtained plans of the trek boats which had been plying passenger services for centuries in Holland.[1] Meanwhile Gilbert was ordered to introduce a passenger service of some kind in 1766, which ran between Worsley, Castlefield at Manchester, and Lymm, with the cost fixed at 1*d* per mile. The only ticket known to have survived is later, and closely resembles the traditional Edmondson-type card railway ticket in its size and green

15 Packet boat ticket

colouring, dating, it is believed, from after 1838.[2] We do not yet know what type of wayleave was issued in earlier days.

Pressure of work obliged the duke to farm out the passenger system to any merchant willing to pay a rental of £60 a year for it, but as no one had the foresight to accept his offer he was obliged to run it as one of the estate's additional services; such were the fruits of ingenuity in an expanding economy that by 1776 the net income in a single year was already £1,356, and by 1800, £4,787.[3] If the first boats were probably cumbersome converted flats of a kind which had long carried passengers on some of the old river navigations, by 1 September 1774 he was running two spanking new packet boats carrying 120 and 80 passengers respectively, though they were sometimes grossly overcrowded.[4] The bows were fitted with sharpened sabres curved like a swan's neck – a threat to the tow lines of slow-witted bargees more symbolic than real. The faster boats with first-class passengers might be drawn by three horses, with a captain, and a liveried postilion up, armed with a curved horn to warn of his approach, spurs and a whip to urge his steeds forward – a gallant and graceful sight, with the boat riding up to 6 m.p.h. on its bow wave and the roof passengers ducking as they sped beneath the bridges. The later Scotch boats could reach speeds of around 12 m.p.h. but this involved working the horses desperately hard and changing them more frequently.

The word packet, which derives from the bundle of official despatches sent overseas by ship, had been in use since Tudor times or earlier. From it was derived the term packet boat, meaning the vessel carrying the royal mail. It is not established that the duke's packet boats were entrusted with this duty, but it would seem probable, since they formed by far the safest and most reliable service between the two towns and Liverpool, with coach links for Chester and Warrington also. Ticket offices were established, of which only the Packet House at Worsley, with its fine stone steps, remains, but the Packet Boat inns at Eccles and Altrincham imply that local agents were licensed as retailers all along the cut. At first the duke was advertising enticingly – 'N.B. Tea and cakes elegantly served for breakfast and in the afternoon in each boat' – but later a French visitor compared the cuisine aboard favourably with the best London hotels, and that was a service which the stage coaches could not rival.

By 1788 Lewis's *Directory* was giving detailed timetables and coaching connections for 'Two elegant Passage Boats for Passengers and their luggage only', leaving Manchester Castlefield at eight o'clock, and Runcorn between eight and nine. The Manchester boat was through Altrincham by ten, Lymm by eleven, and close to Warrington by 1 p.m., where a Liverpool coach waited for it. By 2.30 p.m. the Chester coach was meeting it to a winding of horns at Preston Brook, and she was warping into Runcorn Quay to meet the Liverpool boat on the estuary by 4.30 in the afternoon – just under nine hours or so.

The Runcorn ship also sailed about 8.00 a.m., probably from below the locks, and was picking up Chester passengers from Preston Brook by eleven and the Liverpool coach travellers by 1.00 p.m. – by 2.30 they were cantering into Lymm, Altrincham by four, and so to Manchester Castlefield by six o'clock. The number of canalside stables increased as the service was speeded up. Although stage coaches could occasionally average 8 to 10 m.p.h. on exceptionally good surfaces (the Wild West vision of galloping coaches being almost entirely mythological), the duke's passage boats offered a far safer and smoother ride which no highwayman or footpad could rob, and without the time-wasting halts for refreshment at wayside inns.

A third boat ran daily between Manchester and Worsley, while on Saturday the Manchester to Runcorn ship did not sail until four in the afternoon, and a somewhat hilarious vessel she must have been, plunging her merry passengers through the darkness after a good day at the markets or the races.

The Brooke challenge

As the duke's waterway passed south of Warrington it entered the demesne of Sir Richard Brooke of Norton Priory, who had been Sheriff of Cheshire in 1752. The Brookes were landowners and navigators of the old persuasion – the upper Mersey estuary with its fish garths was their bailiwick, and they had attempted to open the little river Dane, a tributary of the Weaver, to barge traffic.[5] They held two fisheries in the Mersey and maintained that no man was entitled to set foot within their territory, much less dig unsightly ditches for the use of boatmen who would inevitably poach their game, rabbits, wildfowl and fish. Moreover Sir Richard had recently spent some £20,000 on rebuilding the Priory from its foundations, and laying out magnificently landscaped grounds and gardens, which included lakes, so this was clearly no place to bring a canal without a grave infringement of private property.

Well before applying to Parliament to extend his canal to the estuary, the duke had written to landowners personally to obtain their consent, but Sir Richard's reply established his opposition from the start:

My Lord,

I had acknowledged the Honour of your Letter sooner had I not waited to see the Plan of your intended Canal and duely to consider the advantages & Disadvantages of so great an undertaking to the Country, which I have now done, and am fully convinced that the Inconveniences which will accrue both to the Landowners and Tenants will be very great, and as we have already a free Navigation fully to answer all our Occasions a new one cannot be of much Service, therefore hope your Grace will excuse my consent & Assistance to your present Design. I beg leave to acquaint you from my own Remarks and the Declaration of the oldest and most experienced Watermen upon an impartial Examination that at Nip Tides you will not have a sufficient Quantity of Water to carry on your Boats from the Hempstone to Runcorne.

I am with great Respect

Your Grace's most obedient Humble Servt;

Norton, Novr. 23rd 1761 Richd. Brooke

P.S. I have Coal now laid down within half a Mile of my own House for 3^{d}: $\frac{1}{4}$ per Hundred.

Success, or even its prospect, breeds fear and jealousy, and the Cheshire squires, supported by the Stanleys, were determined to oppose canals which might indeed be beneficial to the public but endangered trade and employment on the old navigations, and extended the power of the Gower–Egerton hegemony from Ordsall on the Irwell in Salford, through Tatton, Trentham and Lilleshall to the very banks of the Severn. Something of their fury is reflected in one of the duke's staccato battle orders to John Gilbert from Ashridge on 16 June 1769:

Mr Gilbert,

Enclos'd I send you a copy of a letter and paper I receiv'd from Mr Ligh [*sic*] last past, I think it much better you should take no notice of the aspersion against you in it, (as it is impossible for anyone who is concerned for me to escape the censure of a *Cheshire Gentleman*). On reading the Claims concerning Bridges I apprehend the Commissioners have no powers to order whether the Bridge shall be made over or under the Canal, but that they shall see, that it is executed in a proper, workmanlike manner, and of dimension sufficient to admit of the largest and highest top load. By my answer to Mr Legh[6] you will find I think it wise to hold the Candle – whether that will do I do not pretend to know.

I am Yours most sincerely, Bridgewater[7]

Some historians have found it difficult to see how the duke could have administered his canal and mining enterprise when he visited Worsley so infrequently. The answer is that, like any other large business, the branch estates were run from the centre – the Old Gatehouse of the monastery at Ashridge in Hertfordshire – but it was only in matters of considerable moment that Francis abandoned his habit of dictating letters to Thomas Gilbert and wrote, as he had here, direct to one of his agents. The duke had learnt, as Bradshaw could not do after him, the art of delegating authority – restricting his instructions to policy rather than to detail, leaving a generous leeway to the men to whom he had entrusted his affairs, and backing them with his fullest support in emergencies.

Tomkinson, the duke's Manchester lawyer, and also one of his stewards, was driven almost frantic by the Brooke opposition. By 1770 he was recording the opening of negotiations in words which read like a medieval challenge to combat, with a special summons of sixty folios for the Canal Commissioners to sign before issuing their warrant asking the High Sheriff of Cheshire to summon the jury which would assess 'the value of certain pieces of land in Norton to be cut for the Duke's canal.'[8] The commissioners announced their meetings to settle these disputes in the local journals, before gathering at the Old Coffee

House in Manchester, but the juries were usually empanelled from men with some knowledge of the area under dispute. Francis was obliged to pay all expenses if he lost, and by the time these assemblies and jurors had wined and dined at his Grace's expense the bill amounted, as Sir Richard well knew, to no small sum.

In one matter the Brookes were convinced that they were secure. When the Trent and Mersey Act was passed in 1766 it contained a clause that the duke's cut would never be allowed to pass within 360 yards of Norton Priory, and that no unsightly piles of 'Gravel Sand Clay or Earth' should be allowed to lie within 500 yards of the house for longer than a year. They stood entrenched behind this clause with all the power of law and the sanction of Parliament, but Francis was faced with an engineering problem. His original plan to take his cut to the Hempstones, on the south bank of the Mersey, had been abandoned when he agreed to link with the Trent and Mersey at Preston Brook; but now the rising contours of Windmill Hill, combined with problems of water supply and depth, made it, he claimed, impossible for him – 'extremely difficult' scored out in his own hand, and 'absolutely Impracticable' substituted – 'to execute under the restrictions therein directed and prescribed'.[9] This he submitted in a petition to Parliament, which included the weak plea that he had 'not been sufficiently Informed' of the particular Restricitons . . . at the time of the passing of the said Act'.[10]

Parliament was not amused. It did not meet to retract Acts which had been passed only four years earlier, and the Commons debated the matter hotly. A Mr Seymour admitted that 'the Duke is a truly great man', but added that the 'father of canal navigations' should not become their destroyer by, as he hinted, making a mockery of law. Sir William Meredith thought that 'the public must go unaccommodated rather than private property should be invaded'. Lord Strange was, as usual, adamant in his opposition to the duke and anxious that the matter should be decided by a committee, but Charles James Fox ably opposed that idea. Eventually a Mr Stanley rose to say that the Act was not so very unreasonable as long as the canal did not run too close to Norton Hall, and Robert Wood, the duke's old tutor, replied, 'I cannot say what the Duke of Bridgewater's intentions are; but I think he will stop here.'[11]

Though Sir Richard's supporters lost the motion to put the matter to committee by thirty-three votes, it was a costly victory for Francis. In a Parliament composed of many of the major landowners in the realm, he had set a precedent by which the new arterial canals could be carried almost into a man's private garden, and, as he soon found, it was one thing to get it passed, quite another to enforce it against such powerful and angry opponents. Early in July 1771 Henry Tomkinson, the duke's Manchester solicitor, rode to Norton to pay Sir Richard £315 for a small parcel of land which even Brooke's men had valued at only £268; in another transaction Tomkinson raised the agonised cry 'But the terms insisted on by Sir Richard were so high, that it was thought prudent to pay this sum on account, that the workmen might proceed in cutting in this estate, which adjoined Preston.'[12] Sometimes John Gilbert rode over with Tomkinson and his clerk, but tension was increasing, while the duke lost more desperately needed income – by 20 October 1771 the profit from his canal amounted to only £3,546, while his debt had swollen to £133,219, on which he was paying £5,400 interest annually.[13]

The duke and Gilbert would not take such a challenge lightly. In January 1768 Francis had bought land near Salthouse dock, Liverpool, for what would later become the duke's dock,[14] and when the cut reached Lumbrook, near Lymm, in May 1770 they used an outflanking movement. Goods were unshipped at the canal head, carted to Warrington Quay by Thomas Longshaw, the main transport contractor, and reloaded on to the growing fleet of flats which the duke had been building at Bangor on Dee for the thriving trade on the Mersey and Irwell Navigation and in the estuary.[15] By 1774 at least Worsley coal was being sold in Liverpool,[16] although it would have been transshipped to the river flats at Barton, or taken down the Gut at Cornbrook, rather than carried via Lymm. Meanwhile the duke built himself a small house in the classical style at Runcorn – still used by the Bridgewater Department – to serve as headquarters and main offices for his campaign.

Since Brindley was almost always away on his many canal projects, John Gilbert was obliged to draw all the engineering work and supervision into his own hands, even coping with what Wedgwood vividly described as a mutiny among the navigators, but by the time the cut stood at the borders of the Brooke demesne he had the skilled assistance of his son John, who had, with his brother, attended Manchester Grammar School, and served his apprenticeship under Matthew Boulton junior at the Snow Hill and Soho works in Birmingham. In 1765 his father had written to Boulton asking that the lad should be allowed to attend the scientific talks given by James Arden, an itinerant lecturer, so that he had received as sound a training in engineering as the times would allow.[17] Unfortunately the son did not inherit many of his father's gifts. A competent engineer who is said to have introduced new machinery and better methods of gaining coal at his father's mines at Kidsgrove, near Clough Hall in the Potteries, his careless and ill spelt letters[18] reflect something of that quick temper which gained him a reputation as a persecutor of Methodists, pursuing them from their meetings in the village cottages with a hefty club until eventually he met with a superior force which threw him and his men out on their necks.[19] He died in 1812 and was buried in Audley churchyard.

The change of direction from the Hempstones to Runcorn, in addition to the exorbitant fees demanded by Brooke at Norton, added a vast deficit to the duke's overburdened finances, but he began the construction of wharves and warehouses and the great flight of ten locks to carry his flats down the side of the hill to the Mersey at Runcorn. Francis was convinced that these locks would hold water better and be far cheaper to build if they had brick walls ribbed with stone and 'pounded' at the back, but when Brindley was consulted he disagreed violently. 'If these locks stand,' he declared, 'they are the only locks that ever stood without a dry wall!' This was the last of the series of arguments between Francis and his consulting engineer, and Brindley was proved wrong. In after-years the duke would recount the tale over his port, ending triumphantly, if bluntly, 'They may piss, but they've stood!'

Brindley had been dead three months by 1 January 1773, when the flight of ten locks – the only ones on the Bridgewater cut – were completed, and Josiah Wedgwood was ferried across the Mersey to admire their 'Gates, Aqueducts, Cisterns, sluices, bridges, &c &c the whole seems to be the work of Titans, rather than a product of our Pigmy race of beings, & I do not wonder that the Duke is so enamor'd of his handiworks, that he is now in the fourth month of his stay in this place & is expected to divide his time between Runcorn & Trentham for the remainder of the summer'.[20] But though the Brookes were outflanked they resisted so stoutly that the duke toyed with renewing his case before Parliament. His debt stood at £210,115 by 25 December 1773, and Christmas must have been less than happy, with no prospect of settling the Norton dispute. On 29 January 1774 the following note was delivered to the duke: 'Sir Richard Brooke presents his respectfull Compts to his Grace the Duke of Bridgewater, and desires to be inform'd whether he intends to renew his Petition to Parl. this Sessions for Power to alter the Course of his Canal [through his land] from that which is now prescrib'd.' Angrily the duke scribbled his draft reply beneath for Thomas Gilbert to copy. 'The Duke of B. presents his Com: to S.R.B. and informs him that he is come to no resolution about presenting any petition to Parliament either this or any future session. When ever a Petition is presented SRB will from the course of Procedure in Parliament have many opportunities of doing what he thinks proper.'[21] For a year Francis allowed matters to ride, so gaining further support from the merchants whose goods to and from Liverpool were long delayed by being unshipped and broken as they were carted to Warrington Quay, or jogged round the embattled borders of the Brooke estate.

On 17 October 1775 John Gilbert and Henry Tomkinson rode over to Norton to make the Brookes a final offer. It was sternly rejected, and as his canal debt was standing at well over £229,000 Francis was reluctantly obliged to take his case back to Parliament. He met Tomkinson and Thomas Gilbert at Cleveland House in London to co-ordinate their plans, even going so far as to engage John Lee as barrister to plead in the House.

Then, suddenly, Thomas Gilbert told Tomkinson to come urgently to his town house in Queen Street to show him a letter written by Sir Rowland Hill as arbitrator. The Brookes realised that public opinion was swinging against them and had bowed to the inevitable. It was an event important enough for the duke to write to Sir Rowland personally on 29 October 1775:

Sir,
I am much obliged to you for your Interposition in order to consiliate the dispute between Sir Richard Brook and me Relative to the land near Norton, and for the proposals which you have been pleased to send me for that purpose. The sum of two thousand pounds which is expected to be paid by me is a considerable one to be added to the very great Expence I have been put to in lifting water for supplying the latter part of my canal; had the offer come before that work was done, I should have had much less objection to it, and even now I am willing to leave myself entirely to you to accomodate the matter as you please if you find a proper disposition in Sir Richard Brook for that purpose. But from a letter I have lately received from him about a Tree which I bought of him, he takes Notice of our dispute about the Navigation in such terms as do not afford the least prospect of an accomodation, a copy of that part of his letter I send you inclosed, and am, with a just sence of your kind intentions in the trouble which you have taken

Your very Obedient
Humble Servant
Bridgewater

P.S. I write to you in Confidence that no use will be made of it to my disadvantage.[22]

The unfortunate Hill had a thankless task in his arbitrations. On 15 December 1775 he wrote to Thomas Gilbert from Hawkstone: 'On receipt of yours I communicated the Contents of it to Sir R. B. and told him that if he persisted in his Demand and would not leave the matter to me, I should be obliged to give it up and let him fight it out with the Duke in the best manner he could'. There was grave concern lest the canal might dissect 'the other Springs which run through or rise in the Stews and pass through his water Closet in the House.' The reason, as Hill again pointed out, for 'Sir Richard's backwardness in not coming into the proposals in bringing the cut so near his House, is owing to his fears lest his friends should think he has been influenced by pecuniary motives'. In retrospect this reason is clearly inadequate, and though it is true that the duke had not, in this case, done his homework properly, Hill's sympathies lay more with him than with Brooke. But at last his tactful perseverance brought a compromise.

Work began at once, and by 21 March 1776 the pipe dream which Francis Egerton had conceived seventeen years before in the face of scepticism and derision had at last come true. Manchester had been joined to Liverpool by the Duke of Bridgewater's canal, and 600 navigators feasted on roast oxen and drank the duke's health at his expense.

By 1788 John Aikin could write that Lancashire had distinguished itself 'especially of late years, beyond any other [county] in the kingdom', as a commercial and manufacturing centre. 'Manchester,' he wrote, 'which has long been noted for various branches of the linen, silk and cotton manufactory, is now principally conspicuous as the centre of the cotton trade, an immense business, extending in some or other of its operations from Furness . . . to Derby, north and south; and from Halifax to Liverpool, east and west.'[23] Of Liverpool he could write that it had become unquestionably the second port of the kingdom, superseding Bristol, the reason being 'the superiority of the canal navigations . . . above the difficult and uncertain navigation of the Severn'.[24] The canal idea had been more than justified. Sir Richard Brooke died five years after the canal was opened, and was buried with his ancestors in Runcorn churchyard. As the moss and grasses crept back along the banks of the canal it became, not the eyesore that the local squires had feared, but an asset to the landscape.

Notes to this chapter are on p. 192

12

Man with a mission

While the duke remembered, London had long since forgotten the lovely daughters of Squire Gunning and even Kitty Fisher, the graceful courtesan who had reigned after them. The century was growing older, and the duke with it. In a sense the canal had taken seventeen years to build, but the work was never finished, as the network of underground waterways spread deeper into the coal seams under Lancashire. Some twenty years had passed since the young grandee, returning from his European travels, had begun to take an interest in transport and mining. By 1776 he was forty years old, and had changed considerably in character and appearance.

The slim youth who a wry wit had wagered would be blown out of his saddle on a windy day had grown stouter, developing what seemed to his friends some very curious eccentricities. He frequently spoke in broad Lancashire, interlarding his conversation with strings of oaths, even when ladies were present. Apart from his sister Louisa, and his probably brief encounter with Miss Langley, his experiences with women, especially his mother, had been persistently unfortunate; with the passing years he began to eschew their company, surrounding himself with a few well tried friends and cronies. During the Trent and Mersey negotiations Wedgwood found one of these something of an obstruction. 'I had a Conference with his Grace of Bridgewater yesterday morng. for an hour, & hinted the matter to him but the famous Brown was with him, & I could do nothing to purpose. The Duke recd. & conversd. with me just in the same affable, familiar manner as usual when he first began to know us, by *ourselves* . . . & not from his *go betweens.*' There followed a curious exchange between Brown and Wedgwood, who insisted that his life was as much devoted to the ladies as the duke's friend was to noblemen and gentlemen. 'He assured me they were not ungratefull & intimated that I was nearly as famous amongst the Ladys, as he was amongst the Gentn.'[1] In 1777 the duke was employing Lancelot 'Capability' Brown to landscape much of the 4,000 acres of Ashridge Park.

Meanwhile, Francis was held in growing esteem by the people of Lancashire as they realised that he had solved the energy crisis by providing transport for coal, and had contributed so much to reducing unemployment. At Worsley or Runcorn he would be seen at his wharves, supervising the loading of his boats, or checking the accounts, dressed in his quaint coat and breeches of brown worsted made for less than £6 by Baxter, the colliers' tailor – a suit of strangely outmoded cut, fastened at the knees with a pair of old-fashioned silver buckles. The evidence for his personal generosity is now amply established, and more than justifies Dupin's comment, 'The Duke was at the same time liberal and benevolent and must have spent considerable sums in charities, as the universal respect in which his memory is held in Lancashire testifies.'[2] John Benson met him at Trentham, and left a vivid picture of the encounter: 'I had been with my father killing a Buck & I had shot the deer through the head . . . His Grace seeing us take the Buck out of the Park came down to the venison house to see it broke up & seeing the deer was shot through the head he enquired of my father who had shot it . . . & he told my father that it was much superior to the old way of shooting them through the side which His G. said was the way he shot his own . . . he appeared to be about 5 ft 9 in. high . . . the most florid complexion I ever saw in a man, but not from the effects of Liquor, but such as we formerly saw in a fine, stout, healthy farmer and his Hat was a three corner'd one'.[3]

Though the gatehouse at Ashridge in Hertfordshire, crowned by its worn stone crucifix, remained the administrative centre of his activities, Francis was perpetually travelling around the country, organising his estates in person or by dictated letters, borrowing cash for his projects, and cultivating those whose parliamentary support he needed for his control of the canal vote – a subject in which he had become the acknowledged authority. He made a virtue even of the widely scattered geography of his estates – wool came up from Ashridge to be sold to Smith, the Rochdale cloth merchant, while the drovers' roads were busy with his plodding cattle purchased as far afield as western Scotland, the Isle of Man or Ireland, and finding their way mostly to the London market via Ashridge. Arthur Young paid him a tribute for 'attending, and at a considerable expence, to matters of husbandry in the midst of an undertaking that alone would convey his name with peculiar brilliancy to the latest posterity'.[4] As Chat Moss was reclaimed he grew wheat, barley and potatoes there and used a clover rotation, while some £2,000 worth of hay was cultivated at Worsley annually for mules and horses used on his canals and estates. With characteristic generosity he allowed John Gilbert to farm his own land at Worsley, while Charles Hulbert recorded that the duke 'kept his people busy working on cotton manufacturing, or the canal, or the estate farms' – most probably outwork gathered in to the Brick Hall at Worsley, while labour was clearly interchangeable.[5] His agents scoured the country for bullocks, which the drovers brought to Worsley for fattening and sale on the Manchester market, but cattle bought for John Gilbert were carefully differentiated from those belonging to the estate.[6]

As gout began to catch him the duke used his carriage or his packet boats more, but generally took a saddle horse or two along on a loose rein to inspect canals or factories of interest on his itineraries, on one occasion even travelling all through a Christmas day to complete some urgent work. His agents also found themselves summoned away at the most inconvenient times to attend to some new crisis. Their employer preferred to be the caller when he had business to transact, saying, 'If I go to them I can leave when I please, but if they come to me they leave when *they* please,'[7] and he was 'punctual in his appointments, precise in his arrangements, and economical both of time and money'. Although his canal debt had risen to £280,000 by 1777,[8] when his mother died, the waterway was by then almost assured of success, and, as he knew how to borrow, the future prosperity of the enterprise was reasonably certain; with some £60,000 from her estate his worries were reduced. So after her death he paid Timothy Caswell, a distant relation, £2,000 a year to maintain her house, Cleveland Row – later Bridgewater House – for his use: 'he always came, and was at his desire, treated as Caswell's guest; though he was consulted both as to the dinner and the Company.' It was there that the duke probably met Bradshaw, to whose care he would entrust the Bridgewater system when he died.[9]

But it was not sufficient merely to build a canal – adequate dock facilities at Liverpool formed an equally vital part of the system if the duke's extensive carrying fleet was to be kept busy. The Mersey and Irwell had a dock at Liverpool, and the duke also needed one for his exports and imports awaiting lading to or from the ocean-going ships. Exactly when Francis first obtained land for his Liverpool operations is not known precisely, but by 7 November 1776 Thomas Gilbert was writing to John with the duke's latest instructions in what would prove one of the knottiest problems ever faced in his struggle to open Manchester to the sea:

> Dear Brother – I am favoured with yours and have this day shewed it to the Duke who would have no further application made to the Corporation of Liverpool concerning the request he made for a new Lease, but thinks the application should be made for a renewal of the Lease . . . how unreasonable the Corporation have been in rejecting the Duke's request for an extent [extension] of his ground –
>
> The Duke approves very much of the manner in which you mean to employ the hands for finishing the work they are about at Runcorn & hopes you will get all the shallows in the Navigation deepened before the Spring – he says, there was a Perch of the Sutton Lime ordered to be sent for him to London – some time since, but not having heard of its arrival, is afraid it is hasting to decay – to know by what method it be sent & when – I am with best respects to the children Your Affect Bro. &c. The Duke hopes the Mayor of Liverpool will give the necessary protection to the Boatmen, but if any further are wanted you might let me know [a probable reference to press-ganging].
>
> I think the spirited preparation we have made for war will discourage France & Spain from any attempts against us, and we are extremely well prepared to defend ourselves & annoy them if they wish it.[10]

The prophecy was not fulfilled: in 1778 France joined with those

Americans who were fighting for independence, while the following year Spain added its forces to the alliance against Britain. The duke's implementation of the arterial canal idea had come none too soon, to steel the national economy against the coming ordeal; his policy was then more than justified, contrasting favourably with the obduracy of the Old Navigators, the packhorse interests and the 'Cheshire Gentlemen'.

Wherever the canals were cut existing industries were better served and new enterprises began to flourish. A typical example was the Gilberts' salt mine at Marston, which used the Trent and Mersey and the Bridgewater to carry at least 12,000 tons a year, the duty being assessed by the Salt Officer, for whom the duke had built an office at Preston Brook quay at his own cost.[11] Another example of the duke's liberality was the assistance given to the Gilberts' black lead factory at Worsley, which was also located close to the canal transport system, described in a letter from Thomas to his brother –

Navistock, 29 June 1778. Dear Brother, I am now with the Duke of Bridgewater at Lord Waldegrave's, and have by his Grace's directions to tell you that he proposes seeing you in 10 days time, but will give you notice some days before. If Lord Gower goes to Trentham before that time, which seems probable, his Grace means to stay 2 days there & will give you notice to meet him, which will give you an opportunity of conferring with them both on Lord Carlisle's affair.

The Duke is so obliging as to permit his people at Worsley to make an engine for pounding the Black lead, as you desire, the sooner it can be finished and sent the better.

With regard to the application, his Grace thinks, from the situation of his Dock, at this time, it may be proper to apply once more to the corporation, to know if they will allow the extension in the manner he has requested it, & he is then rather induced to renew the request, as he hears they have allowed the like priviledge to Mr Raburn [*sic*].

His Grace would have you immediately discharge the 2 Mr Banks's – as to obtaining a Land place for the young man, that is not an easy thing to get – it would require both time & good Interest to accomplish – he will talk with you upon it when he sees you; but would not have you delay the removing of them on that account as he is very sensible how much his affairs suffer under their present management at Liverpool.

My wife, thank God, continues recovering fast & I hope we shall get out of

Town soon enough for my attending the next Navigation meeting on the 29 July –

I shall stay here for a few days & then return to London & go from thence to Ashridge, for a few days, before his Grace goes for Lancashire.

I desire my best respects to sister & nephew Your ever Affect. Bro'
Thomas Gilbert

I'm glad to hear so good an account of our Black Lead.[12]

Cheshire salt was an immensely valuable commodity in an age lacking refrigeration – it was exported in considerable quantities and used for preserving meat through the winter, and on the ships sailing from the growing port of Liverpool. China was also becoming a major export – between 1778 and 1782 21,745 large crates of china were carried on the duke's boats alone from Preston Brook to Liverpool at 7½*d* a crate[13] – the relationship between the duke and the Trent and Mersey Navigation was always very close, and though at one stage he treated it too much as though it was his own waterway, to the aggravation of some of the merchants trading on it, he also lent it skilled staff and labour, such as Thomas Wallwork, who had superintended the building of the basin at the duke's dock at Liverpool and was engineer in charge of repairs on the Bridgewater.[14] Vast warehouses were required at all his depots, and towards the close of the century Francis was employing twenty porters at Preston Brook alone.[15]

The duke's dock

Meanwhile the battle for extending the duke's docks was faring ill. The younger Thomas Gilbert wrote to his father, John, in 1776 from Liverpool with encouraging news that a Mr Rathbone had the approval of a committee for his plans to extend his premises 35 yards, and that this was expected to be confirmed the next council day.[16] But two years later John Gilbert was writing to his son with a note of asperity: 'Dear Son, the Duke of B. is nowhere & has been informed its doubtful whether the Corporation will grant his Request; for which reason he is sorry I have wrote to the Mayor (crossed out –

notwithstanding my Brother's letter to me) and says nothing would have induced him to have given the Corporation further trouble but the Council granting Mr Rathbone liberty to extend his yard towards the sea. . . . The enclosed letter his Grace desires you will show to the Mayor & if he (the Mayor) thinks it will not be generally approved he will be much obliged to him if he will suppress it – Compliments to the Mayor, his Lady & Family, Your ever Affectionate Father, John Gilbert.'[17]

As the years passed the dock became patently too small for the vastly increased traffic, and the corporation adamant that they would contenance no further intrusion from the Manchester end of the county. By 1789 Robins, the Liverpool agent, was complaining, 'This Morning we had 4 flats to sail out of his Grace's docks & the wind nearly in the same point as upon the 25th inst. and had near 30 men to get them out and only one got clear without running against Rathstone wall – the other three Stroke against it but not much Damag'd – but much of the flood was spent before [they] got off and I am not Certain whether she would all get up to Runcorn in the same tide.'[18]

By 1790 the duke's patience and courtesy was wearing thin and he wrote bluntly to the mayor from Worsley that he was determined to take the matter to a jury: 'Mr Mayor, Since the application which I made to the Corporation in 1776 for the extension of my premises which was refused me as not consistent with the safety of the Port of Liverpool and which at that time was refused Mr Rathbones for the like reasons – but afterwards it being granted to him, I applied a second time to Mr Birch, the then Mayor, but from that application I received no answer.'[19] This letter surviving only in rough draft, the final version may not have been so forthright, but the corporation had already got wind that the duke's patience was wearing thin and, realising that their case was weak, had written the day before to request that a deputation should visit Francis at Worsley, since, as they put it, 'they entertain a due sense of the number & importance of the advantages which this Town & Neighbourhood derive from that Extensive & well directed Spirit of Enterprise & Liberality by which your Grace has been so Distinguished'.[20]

Despite these pious sentiments there were at least two reasons why the Gower–Egerton interest was unpopular with some Liverpool merchants. From 1774–76, before the Leeds and Liverpool Canal took over, the duke was selling Worsley coal at Liverpool with his invariable stipulation that the poor people should be served before the factories were supplied. With the desperate fuel shortage there was sometimes not enough coal available, and the rich were obliged to go 'empty away' or their factories were rationed. Popular ballads were written praising the duke and ridiculing the tradesmen.[21] But more than this, the reputation of Liverpool as a major slave-trading port made its leading citizens hypersensitive to any form of criticism. And there was plenty of that. 'It is to be lamented,' wrote Aikin in 1788, 'that one of . . . [Liverpool's] principal branches is the inhuman traffic for slaves on the coast of Guinea.'[22] The slaving lobby could have been concerned with the intrusion of a man noted for his popularity with the poor and his eccentric bent for conversing with them and providing them with liberal employment. Again, the Stanleys were powerfully near, and may have opposed this extension of Egerton interests.

We have seen that one of the duke's favourite sayings was that 'A navigation should have coals at the heels of it', meaning that the persistent and increasing demand for fuel to provide energy and heating was the sheet anchor of the canal transport system; yet the account books clearly show that it was the carrying trade in general merchandise which provided the icing on the cake and made the Bridgewater an exceptionally prosperous concern early in the following century. Meanwhile the duke's finances could best be summarised in the later words of Joseph Brotherton, a Victorian MP for Salford: 'My riches consist not in the abundance of my possessions, but in the fewness of my wants.'[23] The myth that the duke made a vast fortune within a few years of beginning to build his cut dies hard, while successive writers continue to reject evidence already quoted. In January 1782, for example, Tomkinson entered in his accounts that he had visited a Mrs Mapletoft of Chester 'to prevail on her to wait for her £500 'till Whitsuntide when Mr Gilbert would receive your Grace's rents, and with some difficulty I prevail'd on her to wait 'till Lady Day'.[24] Like any other logical enterprise the Worsley canal estate was expected to pay its own way and not depend indefinitely on charity from the duke's other manors and undertakings, however well they might be doing. Since Thomas Gilbert was land agent for only six out of twelve of the Bridgewater estates – 'his Grace's Shropshire, Northampton, Bucks, Herts, Durham and Yorkshire estates'[25] – with a general oversight of his brother's work on the canal system, Francis alone knew exactly what resources he had at his command. As he studied his accounts in 1782 he must have been encouraged by the realisation that his great gamble was at last destined to succeed. True, his net profit on coal, with its crippling parliamentary limitation, was only £2,000 in 1781, but the carrying trade was bringing in a clear profit of over £7,000 – the goods of Manchester were finding their way to the sea at last in considerable quantities, though the debt had risen another £9,000 to a total of £319,927.[26] Already there were tell-tale expenses for maintenance at one end of the undertaking, while at the other work was still proceeding on improving the dock and warehouse at Liverpool.

It follows logically that the year in which the Bridgewater canal system was established as a success was 1783, when Francis himself decided that the worst was over, and ordered a bonus of a guinea to each collier, and half that sum for every drawer who had been in his service for a year or more.[27] The debt accumulated in paying for these works therefore resembled a pyramid. From 1757 to 1786 it was climbing towards a pinnacle, but in that year Thomas Kent, the chief accountant, pressing hard on his quill, entered proudly in the ledger, 'Debt decreased this year £434–7–7.'[28] After 1786 it was possible to begin redeeming the bonds of those who had lent money to the project as profits began to rise. By 1792 the general public had cottoned on to the profit potential of the summit canals which the duke's example had publicised, and the 'canal mania' sent them hell-for-leather investing in or constructing new waterways – in marked contrast to the difficulties encountered by Francis and his agents when raising his capital.

Between 1789 and 1791 the duke was building his new dock at Runcorn, and when it opened in November he gave a feast for the workmen which rivalled the opening of the locks eighteen years before. An ox was roasted whole and an extra 1,010 lb of meat provided, with 93 loaves and 200 lemons for flavouring, while pepper pots and spoons were handed out free to the navigators. It could be argued that it had taken thirty-two years to complete the two main lines of the Bridgewater canal.

Notes to this chapter are on p. 193

13

Direct rule

As the gout overtook him the duke began to look and behave like a Rowlandson caricature of a squire of the old school. His accent grew broader and his girth expanded proportionately. He washed infrequently, his language was appalling, and he became ever more opinionated and didactic. As the initiator of the English arterial canal system, in which so many were anxious to invest, he was as famous and sought after as any man in the land, but he remained shy at heart and made few new friends. His dislike of women was reciprocated, and, though assured of an understanding welcome from Lord Gower at Trentham, the lady of the house who had succeeded Louisa found it difficult to conceal her detestation of her disreputable old guest; she wrote acidly to Granville Leveson-Gower, who was serving in the navy, 'The Duke of Bridgewater arrived here two days ago as great a treat as ever, and a good deal more indolent, for I do not believe that his Grace's face has undergone the operation of washing these last two Months.'[1]

By the time Francis had been there a week or two she was even crosser with him, and wrote again, 'His Grace of Bridge is with us, not *less* positive nor *less* prejudiced than usual. It is a great Disadvantage to live with our Inferiors either in Situation or Understanding. Self-Sufficiency is the natural Consequence, with many attendant Evils; but his want of Religion makes him an Object of Pity. I do not mean that he does not believe in God, but there he is with the Gout and a disorder in his Stomach, and Death and Immortality never occupy either his Thoughts or Words, and he Swears!'[2]

Certainly the ageing duke was no ladies' man. Shrewd, eccentric and already in his anecdotage, he was as dogmatic as Dr Johnson, whom he much resembled, and like Johnson he stank. But children seemed to like him, and as the duke sped on from Trentham young Susan was up for breakfast to see him safe into his carriage. 'She was,' wrote the exasperated Lady Gower loftily, 'the only person in the family that pay'd him this proof of affection.'[3] Had Lady Gower cared to search beneath the crusty exterior of the man's character she would have found traces of the boy who had tried to learn music in Lyons, and had begun his famous collection of paintings in Rome. From the persecution of his childhood he retained sufficient sensitivity to dislike seeing hares hunted – a rare sentiment for an eighteenth-century landowner – though he did not object to foxhunting. His defects lay mostly on the surface, and Lady Gower had missed the main point in his achievement, for while others had mouthed charity he had done something practical about it. The understandable opposition to his invasion of private property, and the somewhat irrational antagonism of the Liverpool Corporation, make the duke, as father of the canal idea, something of a pathfinder. He also treated his employees – staff and labour alike – with exceptional generosity. The workmen were feasted and assisted with a proper sense of *noblesse oblige*; the Gilberts were not stinted when they asked for help with their blacklead factory; John was given the Clough Hall estate at Harecastle as a reward for his services.[4] However opinionated and difficult he may have become, and very considerable power corrupts most men, he gave up personal comforts to see the canal idea through, and wherever summit waterways went the price of coal was almost halved. We have only to translate this saving into modern terms to see how he managed to conquer the energy crisis of the eighteenth-century, which had held up the growth of industry and population alike. During the Reign of Terror, when the blood of the French aristocracy was dripping from the guillotine, Francis received a singular compliment. His coach drew into Manchester direct from the south with tired horses. Like royalty, he was cheered down the streets, the horses were unhitched and the coach was dragged by hand all the way to Worsley Brick Hall.

Secure in such an impressive trust, he was not greatly concerned if Lady Gower and her friends did regard him as a malodorous, self-sufficient and egocentric old man.

The theory that the duke was a mere cypher in the hands of capable land agents and engineers is no longer tenable. The vast and scattered Bridgewater estates were known only to himself in their entirety, and all the major decisions of policy and lines of guidance stemmed from himself and from his discussions with Lord Gower, and others whom he chose to consult. The Rev. F. H. Egerton, one of his heirs, was often in the duke's company during the closing years, and noted the curious workings of his mind. Egerton would sometimes have to ask the same question several times before getting an answer, and one day remonstrated with the duke, who insisted that he had *heard* what had been said but was considering all the varied computations of the problem before he pronounced his verdict. In his cautious ability to look at all sides of even the simplest problem lies the key to an essentially creative, powerful and reflective mind. When he settled down over his claret his anecdotes – particularly those about Brindley – could be extremely witty, while he retained a fair stock of tales which he would recount in Lancashire dialect.

Mersey and Irwell revival

With the rising canal mania of 1792 Francis once again found himself threatened by the Mersey & Irwell Navigation Company. When the Bridgewater was opened to through traffic in 1776 the Old Navigators, in despair, had offered their waterway to the duke for a mere £13,000, but Francis had insisted that he had set out to break a monopoly, not to establish one.[5] Endowed with a new and more vigorous board, the Old Navigators were considering loading coal from the Pendleton collieries, near Salford, direct into seagoing vessels for export overseas[6] – the first distant echo of a ship canal to Manchester – but they knew that they could not act without the duke's consent. He held a varying number of shares in the Salford Quay concern and, in 1798, five shares in the Mersey & Irwell company – just enough to garner information.[7] The duke's method of bargaining generally consisted of most emphatically refusing any application which might prove disadvantageous, following it with a varying period of reflection, and finally agreeing rather reluctantly if he felt that the proposition was one likely to be genuinely beneficial to the public interest, even if it might sometimes conflict in some measure with his own.

The Old Navigators put their Act for improvement to Parliament and made one of their board, a Mr Fox, responsible for 'negociating the best terms that could be made with his Grace'.[8] Following a meeting at Bridgewater House in London at which Colonel Egerton and Colonel Stanley were present, the Act was withdrawn, but by 1793 Francis was beginning to relent. Hugh Henshall, who had completed the Trent and Mersey and the Chesterfield canals, was engaged with Charles McNiven and Matthew Fletcher to carry out a detailed survey. In 1793 the duke was approached again, and said that 'he would not oppose an Act of Parliament to repair the Navigation and improve it'. Francis clearly saw that there would be enough trade for both concerns, and that the canal alone could not provide an adequate service in a rapidly expanding economy. Similarly, when the Rochdale Canal sought permission to lock down into the Bridgewater at Castlefield, the duke bargained hard. He was not anxious that the Rochdale should compete with the Trent and Mersey as a route linking the main ports, and even seems to have toyed with a narrow canal.[9] On the other hand, when the Rochdale finally obtained its Act, the wording suggests that it was at the duke's insistence that it was built as a wide canal. At first he held out for a high compensation toll of 3*s* 8*d* a ton, but eventually agreed to 1*s* 2*d*, which he claimed would be equivalent to his loss on warehousing and wharfage for the goods which would have gone by road, but which he would lose if the Rochdale was built.[10] When the Rochdale was completed by William Jessop in 1804[11] it brought substantial revenue and a supply of clean water which proved one of the best bargains the duke ever drove. By 1839 the Rochdale was able to bypass the Bridgewater by the curious Manchester and Salford Junction canal, which gave them an escape route underneath Manchester, to join the Mersey and Irwell and link up with the Manchester Bolton & Bury's prosperous coal trade,[12] through Salford.

Power corrupts most people, and Francis, though less susceptible to its influence than some, was not unaffected. He disliked the frippery of flower beds around his houses and gave strict instructions against them, but the estate workers at Worsley thought they would soften the landscape by planting a few blooms to brighten the gardens. When the duke arrived he bore down on these offending plants with his stick and savagely cut off their heads. He had grown to expect unquestioning obedience to his every whim, at whatever cost to other people's feelings – the heavy price exacted by wealth, power and achievement.

John Gilbert was twelve years older than Francis, and beginning to hand over some of his responsibilities to his younger son, John. While working on his projects in the Lake District Gilbert had heard of a rich lead and silver mine to the east, under Alston Moor. A lease was applied for on 24 May 1777, and Gilbert drew up plans for a navigable sough – the Nent Force Level – leading out to a canal above ground.[13] The sough was drilled into a solid sheet of basalt which proved a very different proposition to the sandstone at Worsley, each fathom costing £70 to cut, with at least £5,000 spent on drilling before the ore could at last be extracted. Though the project was not a complete failure, the partners, who included the duke's nephew, the Earl of Carlisle, Lord Gower, and two of John Gilbert's three sons, decided to sell out on 26 June 1798 to the London (Quaker) Lead Company, a benevolent concern, which continued to operate this mine with others owned in the area until about 1880, using packhorses for transport above ground.[14]

John Gilbert's death

John Gilbert died on 4 August 1795 at the age of seventy-one and, according to one source, in the arms of the duke, who must have been surprised to have outlived his old friend. Gilbert was buried in the Egerton vault in Eccles Church.[15] It was a similarity of character and outlook that had made the partnership of the duke with Thomas and

John Gilbert such a success – their practical love of utility, energy, perseverance and dedication to a cause. The Rev. F. H. Egerton, who knew John Gilbert well, praised him highly as a man of 'Uncommon Ability, of rare talent, of Vast Genius'. Like all land agents, he was faced with the problem of combining firmness and self-effacement with benevolence. In 1769, for example, the militia ran riot in Worsley and smashed the windows of Ralph Sharples's house, so he appealed to the Constables' accounts for reimbursement, but 'This J. Gilbert thinks an unreasonable charge and not to be paid', states the relevant entry crisply.[16] On the other hand, he, the vicar of Eccles and others subscribed annually to a small school at Roe Green, Worsley, for teaching poor children to read, learn scripture, knit and sew, while he also left Robert Lansdale, his amanuensis, a small legacy in his will.[17] During the struggle for the duke's dock at Liverpool we catch a human glimpse of his elusive character, travelling with Mr Maddock – described as son of a late rector, and one of the town's Dock Masters – 'jumbled together the other day in a coach by mere accident, upon the road to Warrington, Wherein Mr G. after rallying Maddock a little about seeing his jolly face again at Lancaster, said, That if the Corporation really wished to have the matter settled, and to be upon good terms with the Duke, he was very sure the Duke would have no objections to come into any reasonable proposals, and that he was going *to meet his Grace upon the business.*'[18] His eldest son, Robert, held the duke's living of Settrington in Yorkshire, near Malton, from 1775 to his death in 1820, and he had two sons and a daughter.[19] Robert combined his ministry with some oversight and improvement of the duke's Yorkshire property.

It was in 1783 that John Gilbert had bought a moiety of Clough Hall in Staffordshire, and three years later he was able to purchase it all with help from the duke, who treated his agents with considerable generosity and allowed them substantial time for their many enterprises, while still paying them a handsome salary.[20] His son, John, must have been something of a disappointment. Rather bigoted, short-tempered and careless in the letters he wrote, he still possessed a certain amount of engineering ability and judgement, and it was to him that Clough Hall was bequeathed.[21] On 4 June 1791 we find him writing to Mattew Boulton junior from Worsley, 'I have been desired by some of my friends (who are now entering into a Navigation Scheam from Sankey Bridge to Manchester) to enquire the carracter of a Mr Rennie as a Navigation survayor,' and asking for a private report.[22] Around 1800 he was still living in the Brick Hall and undertaking jobs like this for the duke, but clearly he was not up to carrying responsibilities similar to his father's,[23] though he did manage to hold some of his father's firms and mining enterprises together.

More important than either of these Worsley agents was Thomas Gilbert, who died in 1798, aged seventy-eight,[24] for it was he who had been the duke's amanuensis – personal assistant, adviser, friend and travelling companion – through those dark years when the canal idea looked like a forlorn cause. With his town house in Queen Street, Westminster, as well as his family seat at Cotton, his supervision as 'Auditor to Lord Gower'[25] and his work as a reformer in Parliament, in addition to his leading place in the duke's counsels, it is scarcely surprising that he should be described as permanently in a hurry, but besides this he guided and helped his brother in their coal, salt, limestone, shipping and canal carrying enterprises, to name only a few of their network of activities.

But in the last analysis it was the duke himself who determined his own affairs and formulated the policy which created the heavy transport stage of the industrial revolution. Careless of his dress, smoking long clay pipes persistently, and an inveterate snuff taker, swearing like a trooper, distributing largesse and employment while quite regardless of personal comfort, he outlived them all and sailed out into the next century like some battered but still seaworthy old ship of the line.

With Thomas almost gone the duke was at a complete loss. John's son was not up to the job, and Caswell, his London crony, was probably no engineer. So Francis took on not only Thomas's but John's work himself. By 1797 he was writing to Bury from Cleveland Court, 'I send you Inclosed a Plan of the Intended Recevoir at Worsley, and wish you would have a Plan made upon the same Scale with the Tunnel that is now driving, and shewing the Faults that are now Cut, and when you expect to find the Fault that throws in the Worsley Rock, to be upon a seperate Paper; which may be annexed to the Inclosed Plan, which you will return. I am glad to hear they go on faster with the Tunnel than they have done.'[26] A year later he was still hard at it. 'I have Received yours of the Third Instant wherein you state you shall plough the South End of the Higher Level without stopping the Colliery. By which I take it for granted the two Stops are put down altho' you have not mentioned it.'

'I Inclose you the Plan of The Stone Clough Coal, which I desire you will Fill up and return it to me in your next. P.S. I this morning Received Mr Sothern's letter Dated the 4th Instant & you Will acquaint Him that I approve of the Particulars & Intended Regulations Respecting the Passage Boats &c. mentioned therein.'[27] In our own age of declining standards of craftsmanship it may also be some comfort to know that even the eighteenth-century was not altogether immune, for he issued instructions to Varey, his Cashier, 'to wait upon Mr Phillips and let him know I am sorry for the Trouble he has had relating to The Piece of Handkerchiefs Received from Him, and which I made use of yesterday. I find them in Substance not above one Fourth of those which I first had from Him, and that one Handkerchief which has been in use Twelve Years is now Better than the whole piece last Received.'[28]

Charles Hulbert, who grew up at his uncle's rented farm – Westwoods, at Worsley – left an attractive portrait of the duke and John Gilbert calling in occasionally to 'freely and cheerfully converse' with his tenants, and admire the boy's sketches of Barton aqueduct.[29] 'The sight of the Duke of Bridgewater was always animating to my mind,' he wrote. 'He was the soul of everything in that vicinity; he who could obtain a situation or a farm under His Grace was considered a fortunate man. I see him even now in my imagination, standing on a summer evening in the bow of his Canal Yacht' with Lord Gower and one of the Gilberts beside him. 'The party seemed quite happy . . . for they knew that they enjoyed the honour, love and esteem of every individual within the vicinity; the eye that saw them blessed them. Happy were the inhabitants of Worsley in those days with no want of employment or money.'[30] Hulton described the duke as being above the middle stature, 'corpulent but very active'; he generally wore his own hair, and tied it in the club fashion, with a handsome knot of ribbon over his old brown coat with its large gold buttons.

Hard by the iron foundry and yard at Worsley, and close to the site of the present crossing there, was a swivel bridge where the duke could often be seen watching the work in progress. One day a cheeky girl pushed past him carrying a bundle of loose cotton cardings on her head, coming so close that they stuck all over his hat, hair and coat. The ducal roar of oaths brought half the workmen around him, winnowing him with their hats to get the offending little bits of fluff from his hair, and puffing and blowing, as Hulbert put it, enough to 'set a flat on sail', while the girl turned her head and grinned. But the duke had the last laugh when he got Gilbert to issue an edict that the private bridge would be closed thenceforth to 'all females with burdens'[31] – not an injunction likely to be observed either in the letter or the spirit of the law in eighteenth-century Worsley, which was particularly noted for the gaiety of its wedding, funeral and other parties.

Though it may be hard to equate coal with a magical elixir, it flowed along the arterial canals which the Gowers, Egertons and Gilberts had introduced, and transformed agricultural England into industrial Britain. It was the largest single item carried on the duke's cut, floating straight from mine to market, while the phenomenal population explosion of the old Salford Hundred, which included Manchester, from 1780 to 1900 relates directly to the geography of coal deposits and canal building.[32] Up to 1786 the duke sold in Manchester well below his statutory limit – then at 4*d* a cwt, and from 1793 at 4½*d*, as inflation and consequent wage increases caught the country in their grim grip.[33] Runcorn is a typical vindication of Clapham's theory of the relationship between energy supply, canals and population growth.[34] In 1786 – ten years after the opening of the canal as a through route – there were not more than 800 people in a village noted for its clean air and bathing facilities. By 1801 there were 228 houses and 1,397 people, but forty years later the population stood at 13,207,[35] and a contemporary who knew the duke personally insisted that he did build his cut 'as much to promote the public prosperity, as to increase the wealth of the . . . heirs' who would inherit it.[36]

On 1 February 1793 France declared war on England. The younger Pitt had more than once asked to visit Worsley and the industrial areas of Lancashire which were contributing to the sinews of war. Francis always contrived to put him off, saying, 'He will see how rich the country is, and will find out something which will bear additional taxing,' though in Parliament he was described as a 'dedicated friend to the Pitt administration'.[37] The duke was right, for in 1797 Pitt introduced a Bill – almost immediately withdrawn – to tax goods passing on inland waterways, and then the knell of Merry England, the income tax of 1799; but despite his innate detestation of taxes, when the government was in desperate straits to carry on the war in 1796 and appealed for a Loyalty Loan the duke immediately gave £100,000.

The inclined plane

With Thomas Gilbert in his seventies and John dying, the duke's personal administration at Worsley is marked by two important experiments – the inclined plane in the mines, and the less successful canal steamer. By 1795 the underground canals at Worsley extended about fifteen out of the final total of forty-six miles under Walkden Moor, on at least three levels – the 81 ft contour of the surface canal,[38] the lower deep, and the higher level of the ancient sough – but there was a need to reduce bottlenecks in coal extraction, and to repair boats from the higher level without winding them up shafts, so the duke decided to link the two adits.

Work on the plane began in September 1795, and by October 1797 it was completed.[39] As John Gilbert had died early in August, it is reasonable to accept F. H. Egerton's emphatic reminder that 'Of this, as of most of his other great works, the Duke of Bridgewater was himself the planner and contriver'.[40] Aided by Tonge, his head mining engineer, Mugg and Ledbetter, Francis's plans owed a debt to the inclined planes constructed by his old friend 'Wm. Reynolds Esq., a worthy Quaker'[41] on the Ketley canal in 1788, for they often met on the Lilleshall projects, and the Staffordshire canal board meetings, with Gower. These were held at the Crown at Stone with some thirty or more proprietors and commissioners present, to be fortified by ample meals laid on by Mrs Morgan, the landlady.[42]

The plane, which still exists, though now inaccessible, and shattered by rock falls, lies three miles, in a straight line northerly, from the Delph at Worsley, and consists of a double waggonway 19 ft wide on an incline of one in four.[43] At the top were two locks side by side, cut in the solid rock, while overhead the cave soared to 21 ft to give room for the braking wheel. A starvationer loaded with up to twelve tons of coal entered the lock, and the water was drained away until it rested gently on the cradle, which was 30 ft long and 7 ft 4 in. wide,[44] mounted on four solid cast-iron rollers running on cast-iron rails held by sleepers. About 900 tons of coal could be shifted to the Delph level in eight hours for the hearths of Salford and Manchester, the other cradle acting as a counterweight.[45] It was no small feat to have completed such a major operation in two years, blasting out the tunnel with gunpowder, along the line of one of the typical Worsley rock faults, and working the whole vast cave down by hand with wedges and hammers;[46] here again Francis built for posterity rather than immediate gain, knowing that labour had to be subtracted from other projects with a consequent loss of income. Similarly, the open-air canal was lined along its banks with stone to a standard scarcely achieved elsewhere, though this was not completed during the duke's time, being partly the result of the later use of steam tugs.

The duke's steamer experiment was less successful. A Gloucester man, Jonathan Hulls, had patented a tug for drawing ships into harbour in 1736, but the first practical steamboats were launched on the Forth and Clyde canal in Scotland. In June 1793 John Smith's steamboat was tried out on the Sankey Navigation, and by November 1794 the Mersey & Irwell's board meeting minuted that 'A person in Liverpool having produced to the Committee a model of a boat navigated by machinery . . . Mr Wright was directed to pay him five guineas'.[47] The same boat made a voyage up the Bridgewater from Runcorn to Castlefield in 1797, which impressed the duke sufficiently to start him off on experiments of his own. Around this time the American inventor Robert Fulton had managed to get an introduction to Francis through the Duke of Devonshire, and some sources insist, on tenuous oral evidence, that he helped design the Worsley tug, though this is unlikely.[48] More probably the duke himself worked in conjunction with Captain Shanks, R.N., an able inventor who had dredged the Runcorn sandbanks, and whom Francis had helped with designs for a twenty-four-pounder carronade (a short, large-bore cannon).[49]

The canal men, who loved their mules and horses, looked with growing distaste on boats navigated by machinery, but by 1796 the plans were with Bateman & Sherratt, a Salford engineering firm, which started work on the boilers, gearing and shaft, while the hull was built at the Worsley yard by Benjamin Powell, and the paddles by Joshua Laughlin, the duke's foreman millwright.[50] The smokestack was right in the front of the bow; next came a vast domed boiler embedded in bricks, and finally a mountainous array of flywheels, rods and pistons, hiding the unfortunate helmsman.

The duke's men christened the belching monster *Buonaparte*, partly, perhaps, in derision of Fulton, who took service under Napoleon in 1797 and built a steamer for the projected invasion of England which split in half and sank on the Seine. On her maiden voyage *Buonaparte* set off with the high hopes of its patron and inventor, but as it gathered way, to the mingled horror and delight of the spectators, it was suddenly realised that the funnel was too tall, so it had to be hinged in solemn salutation of each passing bridge. With much puffing and splashing she gallantly dragged eight 25 ton starvationers to Manchester, but at only 1 m.p.h., and, as it was feared that the flailing paddles would disturb the puddled clay on the bed of the cut, Shanks abandoned the experiments. In 1801 Lord Dundas commissioned William Symington to build the *Charlotte Dundas*, one of the first really successful canal steamers, and when a model was shown to the duke he was sufficiently impressed to order eight of them. Unfortunately he died before they could be built, and Bradshaw, the Superintendent, fearing unpopularity with the boatmen, cancelled the order. Later used as a pumping engine, and rechristened *Old Nancy*, Sherratt's engine did yeoman service during a bad burst on Brindley's Stretford aqueduct in 1799.

Robert Lansdale wrote, 'I well remember the Steam Tug experiment . . . between the years 1796 and 1799. Captain Shanks (R.N.) from Deptford was at Worsley many weeks preparing it by the Duke's own orders and under his own eye . . . there is a model of it in the office. . . . It would do upon a deep Canal or River, but not where the depth is only 4 ft 6 in.'

'Several adventurers after that came in Mr Bradshaw's time with Ingenious Models, but Mr B always told them that he would'a trot on in the same way as the Duke had done.'[51] He added that on her maiden voyage *Buonaparte* 'drew eight canal boats of 25 tons each, starting at 6 a.m. and reaching Manchester at 2 p.m.'.[52] This suggests that Fulton may have learned more from the duke's steam experiments than he contributed to them.[53]

The historical background against which these experiments were carried out was a critical time for England. By 1797 Napoleon was emerging from the welter of the revolution, France controlled the Dutch fleet, and only a mutinous navy lay between his gathering armies and the English coast. No one was better aware than the old duke that the fighting power and potential of the French nation rested in no small measure on the efficient canal transport system which its leaders had developed during the last 150 years.

Though others might be concerned about his manners and his fixations, the duke cared little enough. Caroline, the last of his eleven sisters and brothers, had died in 1792, and it must have been a matter of some surprise to him that he had managed to survive the family ailment. The death of his chief agent and legal adviser, Thomas Gilbert, in 1798 must have been a severe blow, but he had the pleasant and intelligent company of his nephew, George, Lord Gower (1758–1833), later Duke of Sutherland, who was Louisa's son, to brighten the last years of his life.[54] Francis was staying at Trentham again in 1797, and a fellow guest has left a portrait of him. 'He was every day (as who in that eventful period was not?) very anxious for the arrival of the newspapers and intelligence from London, and when there was no London bag, which was the case on Tuesdays, he called

16 The duke in his old age

it emphatically a *dies non.*'[55] F. H. Egerton once asked him why he would never accept any decorations, such as the Garter or the Bath, to which the duke replied in his blunt way that he would not have any other man's baubles hanging on his chest, but when the article on the underground inclined plane was published in 1800 he did accept the Gold Medal of the (Royal) Society of Arts, voted to him 'as a testimony of the high opinion entertained by the Society of his Grace's execution of this great work, and his wonderful exertions in Inland Navigation'.[56] Benjamin Sothern was in general charge of the building of the inclined plane, but the duke never entrusted him with the same measure of almost unlimited power that he had granted to Gilbert.

In these last years the duke resembled a ruler drawing the reins of his widespread empire into his own hands – and overworking in consequence – but time was running out and he was anxious to leave his estates, investments and enterprises in as perfect order as circumstances would permit. Gone were the days when he laid down broad lines of policy for his agents while himself concentrating on parliamentary and administrative problems. For a few years he became the very engineer in charge of his own increasingly prosperous coal supply and canal carrying network.

Notes to this chapter are on p. 194

14

Death of a duke

If the continuing work on the soughs is excepted, the duke built only one new canal after John Gilbert's death, though his advice and guidance were eagerly sought on many others. He had, for example, been a shareholder of the Grand Junction since its beginning in 1793, but was elected to the committee in 1801 and 1802, to put through a thoroughly businesslike reorganisation of the financial and general administration. Similarly, he was a major landowner on the route of the Ellesmere and Llangollen canal, which obtained its Act in 1793, though it was not completed until just after the duke's death. Despite having had to sell large tracts of this estate during the lean years, he still owned, it was said, the entire town of Ellesmere, while the early prosperity of this beautiful waterway, with its magnificent aqueducts, has been largely attributed to its links with the canalside Bridgewater foundry, and the prospering agriculture of the earldom.

In seeking parliamentary approval for constructing a canal to join his system to the 127-mile-long Leeds and Liverpool the duke again encountered powerful opposition, but this time there could be no question that the man whose example had almost halved the fuel bills of most of the country was undertaking work of public utility. The Leigh canal would give the 'dukers' access to Liverpool without the delays of the tidal estuary of the Mersey, which was only accessible at Runcorn for a limited time each day;[1] The Leigh also formed a vital link in the chain with the new industrial areas being opened to waterway transport around Bradford, Blackburn and Wigan.

Benjamin Sothern surveyed the Leigh branch, and by 3 February 1795 a committee of the Commons was reporting back to the House that it had examined this engineer, and had considered the benefits and public utility of the plans favourably. By 19 February a strongly worded and accusative petition was read from Thomas Mort Froggatt, Lord of the Manor of Astley, alleging that he had been deluded into giving his agreement by the duke's agents, and that it was high time that the whole process of applying for waterway Acts was thoroughly overhauled – a not unreasonable request; but this was followed by a petition in favour from those who stood to gain from the project.[2] Despite three more petitions, including one from the commissioners of the turnpike road, by 14 April General Egerton was carrying a message to the Lords to 'desire' their concurrence; as usual there was no difficulty there, and the Act received the royal assent on 28 April 1795.[3] By early 1799 Sothern was taking over the land of Thomas Green, Sir Nathaniel Duckenfield, John Ratcliffe and others for the new cut, and the work was probably completed by 1800,[4] though it was only in 1821 that Bradshaw was able to link up with the Leeds and Liverpool, and so create a third water route from Manchester to Liverpool.[5]

Meanwhile the combination of energy supply in the form of coal, and of transport by the canals was at last pouring substantial profits into the duke's accounts. By 1803 the gross income amounted to:

On tonnage carried	£48,403
Colliery profits	£24,300
Lime	£ 91
Net profit after deductions	£65,952

and in a single year he was able to reduce his canal debt by £57,832.[6]

Though the duke was far from parsimonious, enjoyed his port and claret, and liked good food, his canal debt still stood at £162,397 in 1803 and his habits of economy were ingrained from the time when he had cut expenditure to the bone. Yet his capital assets were now vast. His judicious investments – especially in waterways – were reaping rich dividends and the value of his many estates was soaring, so that he could count himself a very wealthy man. He gathered a number of servants in the grand manner of his youth, even employing a perfumier called Bourgouis. In 1799 he subscribed a considerable sum to books, including a set of Bibles with a commentary by Bishop Wilson, an act which, had she known of it, might have appeased Lady Gower's concern for his pitiable lack of faith. Some years earlier he had been among the leading subscribers to a work on the science of surveying.[7]

The Bridgewater paintings

By 1801 Francis was pulling down the old monastery at Ashridge with the intention of rebuilding it, and as his debts diminished he toyed with the idea of returning to his early hobby of collecting paintings. He still had most of the pictures which he had acquired in Rome and had proudly refused to sell even at the height of his debts, as well as those his mother had collected. It is said that his ambition to become a great patron of the arts was sparked off again one evening in 1794 when he was dining with Louisa's son, George Gower.[8] He saw a fine painting which his nephew had picked up from a dealer and was so impressed that he cried, 'You must take me to that demmed feller tomorrow!' During the next few years, to the astonishment of the younger generation, who knew nothing of the other side of his character, he showed great good taste and a remarkable gift for picking the finest paintings. He made one judicious purchase after another, until the Orléans collection, gathered by Louis Philippe, Duc d' Orleans (Philippe Egalité), came on the market as a result of the revolution in France.[9]

The duke first heard about the Orleans paintings from his favourite art crony, Michael Bryan, a dealer and collector of considerable acumen, and author of the *Dictionary of Painters*. The Italian part had been mortgaged to Harman's Bank for £40,000 and Bryan, knowing the duke's love of the Italian schools, persuaded him to buy all 305 paintings for £43,000. Bryan then studied and valued them, and came to the conclusion that their market value was about £72,000 – almost double the price that he had just paid for them, acting as agent for the duke, Lord Carlisle and George Gower.[10] Hertford County Record

Office holds bills from dealers for some of the duke's purchases, including a Claude, for £1,890, a 'Wilson . . . thrown in by Mr Boromi, as by Agreement', and Tenier's *Chymist* for £183 15*s* 0*d*.

Francis went through the collection with great care and selected ninety-four of the finest paintings, which he kept – most of them with religious, and a few with classical themes.[11] After biding his time he and his two partners, Gower and Carlisle, held an exhibition of the remaining 211 pictures. These 'left-overs' he sold for no less than £41,000. As his purchase price for the entire Italian collection had been only £43,000, he had managed to acquire ninety-four of the world's finest masters for the sum of £2,000.[12] About that time he was busy restocking the cellars of Bridgewater House with madeira, and it can be safely assumed that, when the results of their extraordinary speculation became clear, he and his partners would have cracked more than one bottle with Bryan. Some of the Bridgewater collection now forms a significant part of the National Gallery of Scotland, on loan from the Sutherland family since 1946.

Though the duke treated the dealers firmly, he was generous to living artists. He was one of the first to recognise the genius of Turner, and when George Gower rejected one of Turner's paintings because he thought the price too high Francis immediately said that he would pay the full amount. He also gave Turner a 250 guinea commission to paint a companion picture to his 'Vandervelde', and the artist, inspired as much by his patron's trust and enthusiasm as by his gold, produced a copy which far outshone the original in realism and breadth of treatment.

To house this sumptuous collection the duke commissioned the architect James Lewis to build a gallery on to the side of Bridgewater House for £1,500, which included 'a very handsome and liberal allowance for attendances and sundry designs'.[13] There, looking now somewhat infirm, he would spend an hour or two a day absorbing the beauty of the paintings he had acquired so cheaply. 'His habits,' wrote Farington in his diary for 1795, 'are to rise between 8 and 9, and to dine at 5 o'clock, from which time 'till 10 o'clock he remains at the dinner table,[14] and though slowly must drink a bottle of wine a day. Port is his wine. He is a shy man and lives with but a few. Lord

Gower, his sister's son, is a great favourite, and is also a shy man. He dines with the Duke two or three times a week. . . . He goes to bed about 11 or 12. . . . General Egerton is frequently with the Duke. . . .'[15] About this time Lord Kenyon congratulated him on the success of his great gamble on canals, but as usual Francis was looking to the future rather than the past. 'Yes,' he replied in his blunt way, 'we shall do well enough if we can steer clear of those demmed tramroads' – a remarkable prophecy which younger men recalled with surprise when, almost half a century later, the railways began to offer a serious challenge.[16]

But the impression of an idle old man taking to luxury and drink could hardly be more mistaken. Inventions still fascinated him, and he would be seen pottering around J. Bramah's 'Patent-Engine, Lock and Water Closet manufactury' at the west end of Piccadilly, where in December 1799 he bought a patent apparatus for a hay-packing press, complete with two pumps, for shipping on the *Triton* to the surprised Finlayson, his agent at Liverpool.[17] Everything he touched seemed to turn to gold, and his last speculation with Woolmers, near Hertford, was no exception. He bought the house and estate from the brewer, Samuel Whitbread, in 1801 for £16,502 10*s* 0*d*, and though Lewis's

17 Woolmers Park House, near Hertford, as rebuilt by the duke

gallery had proved rather damp and not altogether satisfactory he re-engaged him to rebuild Woolmers – still in the classical style which the duke always favoured.[18] By June 1802 John Carrington, a local farmer and diarist, was noting 'the Duke of Bridgewater (having bought Woolmers) this day came their and the Bells was ringing for him',[19] and he noted that by December the duke was adding outbuildings and a substantial new wing.

It was a combination of canals and the supply of drinking water which tempted the duke to launch himself into yet another major project by the purchase of Woolmers. Since about 1779 there had been plans afoot to link the Lee and Stort Navigation to King's Lynn and Norwich by way of the Fens, and a line had been roughly surveyed by John Phillips, who would probably have been known to the duke, as he had worked on Harecastle tunnel.[20] Once again there were problems with an important estate, and a special map was drawn up to avoid the 'private water or pleasure grounds of Lord Howard at Audley End',[21] and also of another estate at Shotgrove,[22] but when the duke died the driving energy to continue this work was gone, so the Lee and Stort remained an uncontinuing waterway.

The other plan was even more ambitious – to offer an alternative and competitive drinking water supply to London in rivalry with Middleton's New River Company, but he met with powerful opposition from that concern, which was growing sufficiently wealthy to protect its own interests.[23] This part of Hertfordshire contained a number of swallow holes, of which the Arkley was one of the most prolific, running under the drive at Woolmers and emerging a few hundred yards away in the park. By February 1802 the *Times* was being 'confidently assured' that the duke was being offered 'Five thousand pounds a year' rent for the water of the Arkley Hole – probably by the New River Company – and, even if this was an exaggerated figure, he had clearly not allowed old age to interfere with his extraordinary speculative gifts.[24]

It was a new appointment which left the duke free for the Woolmers venture. 'The first time I saw Mr Bradshaw was in 1800,' wrote Robert Lansdale.[25] 'I remember the duke bringing him in to the Clerks office (at Worsley) to show him his Business men & incline to believe he was introduced to the Duke by Timothy Caswall Esq., a Commissioner in the Excise & whom the Duke put in as one of the members for Brackley – after Mr [John] Gilbert's death & Mrs G. with her son John had removed from the Hall to Barton House, which was early in 1796, this Mr Caswall had used to come down to Worsley to stay with the Duke a month or more & write his Grace's private letters as H. G. was quite alone for about 4 years 'till Mr Bradshaw came.'[26]

On 28 January 1803 the duke completed a will which ran to sixty-six pages of closely printed type. It carefully tied up the Worsley estate, canals and mines in a trust which gave considerable power to Robert Bradshaw, the successor of Thomas Gilbert as chief agent there, providing him with £2,000 a year and the use of Worsley Brick Hall. Since there was no entail on any of the other thriving properties, his aim was very clearly '*To the intent* that the public may reap from the same those advantages which I hope and trust the plan adopted in this my will is calculated to produce for their benefit'.[27] In other words, he was determined to preserve the Bridgewater's magical combination of energy supply and transport in the hands of a capable and mature administrator, so that it could continue to serve the public and provide employment. At the same time he was fond of his main heir, Louisa's son, George Gower, the future Duke of Sutherland, had enjoyed collecting paintings with him and wanted to ensure that he, the General, Frank Egerton and a few other relatives would benefit as well as the public. Strachan Holme, who was deep in the family's counsels, noted the 'standing grievance of the House of Sutherland' against the Bridgewater Trust because the will had divorced management from ownership – this grievance, he hinted, coloured Lord F. Egerton's brief biography of Francis[28] – but the dying duke had two factors to consider. First, his Worsley enterprise, though largely patriarchal, had already outgrown and would vastly outgrow the dimensions of a mere private estate. Secondly, there was the character of George Gower, a diplomat and country gentleman, ambassador to the court in France until about the time of the Reign of Terror – a spender of money, as time would prove, rather than a maker of it. Thirdly, there was some conflict of interest between the

Sutherlands and the Bridgewater Canal. With those demmed tramroads looming in the background the old man must have considered all aspects of a difficult problem in his dispassionate way, and determined that the Trust was the more just, if not the happiest solution. Most able men consider their heirs less capable than themselves. Though gifted in some ways, Gower was simply not in the same league as his father or his uncle. The Trust may not have been so great an error, for, as F. C. Mather has so ably shown, when railway competition came, the Bridgewater was more than able to ride out the storm.[29] But even Lord Francis Egerton, who, as a beneficiary had suffered from the split between management and ownership, was forced to admit that the duke's main concern in forming the Trust was the service of the public.[30]

A curious twist of fate decreed that the man who had introduced the modern transport system, and had travelled more extensively perhaps than any contemporary landowner, should die as a result of a road accident to his coach, in London.[31] He had even, on one occasion, voyaged from London to Worsley by water, right across England. Though the accident did not kill him, it brought on a severe attack of 'flu, and he had the Rev. Henry John Todd, his chaplain and librarian, to attend him.[32] Todd had recently transferred the duke's splendid library and manuscripts from Ashridge to Bridgewater House. There is no record of the last days, but as he lay dying in his palace in London Francis had ample opportunity to look back over the extraordinary rake's regress of his life, which had followed Voltaire's dubious injunction to love like hell when young and work like the devil when older. Starting with a lost childhood in which he may well have wandered homeless for a while, since the Gowers and the Russels refused him, his *affaire* in France and Wood's patient tolerance, which had fostered his sudden craze for docks and canals, he had progressed with fencing, riding and falling in love with Elizabeth Gunning – all stock activities of any eighteenth-century beau, but then came the serious business of solving the fuel crisis created by the exhaustion of the common and borough woodlands. Summit-level canals were not new, either in Britain or overseas, but people in England needed an example – proof of the canal idea, proof that they would pay, proof that they would not freeze in winter, technical proof on aqueducts and locks – and this was what the duke's great gamble had provided. It had taken over twenty years of hard slogging against adamant opposition to the invasion of private property to convince the nation that the public genuinely stood to benefit from canal transport, and even then there was opposition over extending the Liverpool dock. Fulton was right in saying that only a childless enthusiast would have attempted such an undertaking, particularly in a nation more democratic than France, where every step had to be painfully explained and justified. Hence the Latin inscription on his unostentatious memorial in the family chapel of Little Gaddesden Church near Ashridge records that 'He sent barges across fields the farmer formerly tilled', establishing the principle that public transport benefits took precedence over rights of private property, though such a course inevitably made him many bitter enemies.

It is now a well attested fact that he achieved all this at the same time that he remained one of the most benevolent and charitable men of his age. Including his mines and docks he was one of the major employers of the time, with probably over two thousand people directly on his payroll, and many more working on his scattered estates.[33] Inventors, engineers, authors, artists, architects and the vast number of direct and ancillary services which evolved from his activities, all had cause enough to be grateful, while the public had its fuel bills reduced by varying but substantial quantities. As a Christian Utilitarian, the duke reaped the condemnation of his young cousin F. H. Egerton on theological grounds, but those who quote this statement out of context omit to mention that the book in which it is to be found was written to acknowledge the importance of the duke's contribution to English history.

Egerton both praised the duke's public character highly, and briefly condemned his total subjection to the profit motive by which, as he put it, everything linked with the great Worsley enterprise was considered 'as in a Merchant's Counting House, exclusively upon the calculations of profit and loss, and individual interest'.[34] But this assertion must be seen in the context of early nineteenth-century theological and philosophical ideals. Egerton was part of that new order which criticised the integrity of the East India Company: an ordained member of a reawakening Church, and a Fellow of All Souls, well acquainted with the newer ethic. The old mercantilist order was changing and the energy crisis of the eighteenth-century little remembered or understood by the time he came to write. For the duke an efficient business was a cure for unemployment, and he saw no separation between that and his own estate interests – but he looked after his people with generosity and did have the public interest at heart, for everything in his character attests that he would have scorned to form the Trust for the public benefit if that had not been his genuine purpose. The end product of his thinking was the broad network of canals which his example and tenacity had caused to be engraved on the green hills and fields of England, while he could look back on the Delph at Worsley as the place where the heavy transport stage of the industrial revolution began. Strachan Holme, quoting Meteyard, described him as a 'kind man and a true friend – a little wayward and eccentric in some things; but with an instinctive perception of truth and genius in others'.[35] Miss Edith Malley, who

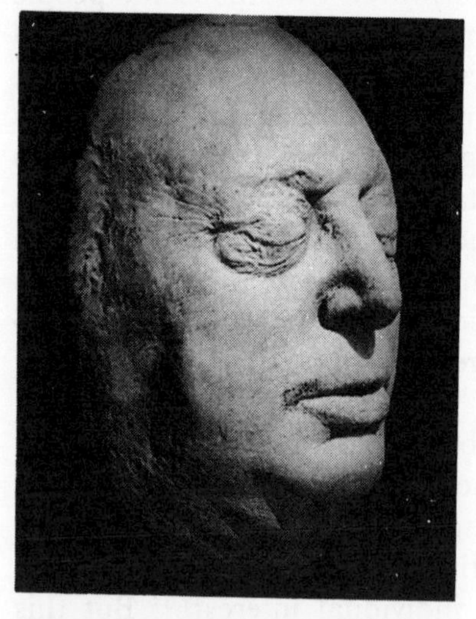

18 The duke's death mask

studied documents now long lost, stated that he was the kindest of employers, and noted the substantial number of 'entries recording the gifts of money to families in sickness or in want' or to men who had been injured.[36] Inevitably, the first man to drive summit canals through private property made powerful enemies who did their utmost to blacken his name, but the myth proliferated by Samuel Smiles that he underpaid his employees has long since been exploded.[37] The more reliable witnesses, Arthur Young, F. H. Egerton and Sir Joseph Banks, were among those who appreciated that the duke's works were his own achievement, and that this man was no cypher in the hands of others, as Thomas Gilbert's letters now so clearly show.

The duke died about 3.30 in the morning of 8 March 1803[38] *The Times* reported that his face wore an expression of placid composure which is reflected in the death mask which it was fashionable to have made in those days.[39] As the body showed no signs of decay, the burial was postponed to give his relations time to foregather. Though he had asked for the simplest possible funeral, the procession ended up with a 'hearse and feathers with six horses; his Grace's carriage with six horses; three mourning coaches with six horses to each; ten outriders and the usual retinue of mutes and other attendants',[40] winding out of early nineteenth-century London to be buried with his forebears in the peaceful churchyard at Little Gaddesden, near Ashridge, in Hertfordshire.

Though he was dead he was not forgotten at Worsley. His personality pervaded the place and the people. The duke's coal still emerged from the duke's mines, the dukers remained the trimmest vessels in the land as they glided under the duke's bridges, and the duke's men still worked the duke's cut, as a few still do to this day.[41] Whenever a knotty business problem loomed up the lesser financiers who followed in his wake asked themselves anxiously what the duke would have done under similar circumstances,[42] and even now the legends are still recounted of his tipping a lad a fourpenny piece, or lending a hand with some job, in the houses that cluster down the duke's cut. He has passed into local lore, and it is even claimed that at midnight, once every year, when the fog swirls over Walkden Moor, a coach drawn by six *coal-black* horses goes clattering through the village with the duke's intent face framed in the window.

Though the dukedom of Bridgewater died out because Scroop could not afford to enable it to collateral branches, the earldom passed to his cousin, Lt.-General John William Egerton, MP for Brackley, who was fifty-nine and childless, and on his death to his brother, the Rev. F. H. Egerton, who, as a Fellow of All Souls and of the Royal Society,[43] was a tolerably accurate, if eccentric, scholar, and author among other works of the *Letter . . . upon Inland Navigation* which very properly condemned the current fashion of giving 'too much credit to Brindley in connection with the making of the Bridgewater Canals and too little to the Duke'.[44] The General inherited the main family seat at Ashridge and the duke's bank balances and investments, amounting altogether, it was said, to some £600,000, and set Wyatt to rebuild the partly dismantled monastery in the Strawberry Hill Gothic style which would have clashed with the duke's classical taste. The will had granted a year's wages to all who had been with the duke two years or more. The General also inherited the Ellesmere lands and Bridgewater estates in Lancashire and Cheshire on condition that they would be returned to the Trust on his death,[45] but Brackley was kept with the canal Trust so that the two pocket borough members, the General and Bradshaw, could keep an eye on canal affairs in Parliament.[46]

Bradshaw and the Trust

In forming his complicated canal Trust, to which he had devoted considerable time and care, the duke was determined that his heirs should receive the full benefit of his lifelong labours, that a competent business man should administer it, and that his own handiwork should in no way perish. Robert Bradshaw was the man whom he had chosen and groomed for this momentuous task, and he even made him take an oath that he would serve the canal idea as its founder had done, without deviation or hesitation.

To this very proper ideal Bradshaw lived true. His character has been the subject of learned controversy between F. C. Mather and Eric Richards,[47] but the fact that the Bridgewater could fight on as a very viable business concern indeed through all the years of railway competition, through times when even Lord Francis Egerton could describe himself despairingly as 'the last canal' which was destined soon to die an honourable death[48] – this was due in no small measure to Bradshaw's dedication to the duke's bequest, and we must not overlook an element of propaganda in the writings of the railway lobby, which went far to blacken his reputation. Bradshaw the man – incapable of delegating authority, and treating his family disgracefully – has to be sifted from Bradshaw the steward, who worked all hours to keep the Bridgewater viable, and gave himself unstintingly in that cause. Unfortunately he had, in his admiration – even imitation – of the duke, failed to acquire all his characteristics, especially that courtesy and common touch which had made his patron popular. It is unthinkable to imagine Francis or John Gilbert complaining, as Bradshaw did, of exile among the 'refractory clerks and rascally flatmen' of Worsley.[49] Such a sentiment, knowing as they did the dedication of men like Robert Lansdale, would simply never have crossed their minds, for they thoroughly enjoyed most of the work relating to their waterway[50] and would have considered any such comment in extremely bad taste. To this must be added the division between the Superintendent and Lord Gower, the beneficiary, who naturally became an absentee landlord, scarcely visiting his Worsley estates for thirty years. In 1825 Gower compromised with the railways. He bought £100,000 worth of shares in the projected Liverpool & Manchester line, which were worth over £73,000 more than he paid for them by the year before Stephenson's track was opened in 1830, but no one of sufficient stature followed the duke as leader of the canal cause, and Gower, who persistently lived above his means, great though they were, would not have been interested in doing so, as the duke had probably realised.

Gower's problem was that his eldest son would inherit the dukedom of Sutherland, while his second boy would inherit the Bridgewater fuel and transport system, which was bringing him in net profits ranging from £50,409 in 1817 to £119,497 in 1824 – staggering sums when it is remembered that even a skilled man's wages still seldom rose above five shillings a day. It followed that he had to devise a policy which would enable the elder son to arrive at some financial parity with the younger, and his chief agent, James Loch, carried through the Sutherland enclosures and clearances which, though inexcusable by modern standards, were certainly no worse ethically and probably a good deal better than many similar evictions.[51] If only Gower and Loch had put the same effort into running the canal transport lobby that they expended on their Highland chimaera they might have made sufficient money to make the Gordon clan territory into a more viable entity than an empty deer forest. Nor were the Gowers much in sympathy with the Highlands – Elizabeth Gordon, Countess of Sutherland (1765–1839), had not grown up in her clan bailiwick but in Edinburgh and London,[52] while Gower actively disliked the sound of the pipes. He also suffered from poor eyesight which made him gradually withdraw from the more onerous burdens of public life.[53]

Apart from the vital link with the Leeds and Liverpool along the Leigh Canal, the reign of Bradshaw at Worsley was largely a holding operation in which the duke's agent liked to delude himself that he was jogging along in the same old way that his patron had done, whereas in vital matters like the introduction of steam tugs he showed his total alienation from that spirit of adventurous engineering and enterprising innovation which had marked every step in the Duke of Bridgewater's achievement. Though Bradshaw kept wages in the mines unacceptably low, the critical fact was that he was there. As the duke had probably foreseen, Gower was an absentee landlord, but Bradshaw ruled in his place – at least there was someone with substantial authority by whom the rights and wrongs of a matter could be settled, and that is what chiefly mattered. Moreover no one has yet questioned Bradshaw's skill as a transport manager, or his business acumen – he was far closer to being the first 'Manchester man' than ever the duke was. The dichotomy between running the canal well and finding vast sums for the Sutherland enclosures and experiments proved a severe strain for Bradshaw, who was also MP for Brackley, while the duke's will had made inadequate provision for repairs and maintenance. By about 1819 he was suffering severely from neuralgia, and gradually became 'rather infirm in mind, and eratic . . . and it was reported officially that he was very much under the influence of two maid servants'.[54] The stage was set for a tragedy which would shake the Trust and the Gower–Egerton interests to their foundations.

Notes to this chapter are on p. 196

15

The aftermath

'Mr Bradshaw was recommended by the Duke to appoint his son . . . as his successor – he had, for some time, assisted him. He, however, appointed Mr James Sothern, one of his Staff.'[1] Behind these few innocuous words of the late librarian at Bridgewater House lies the tragedy of the interior conflict between Trust management divorced from ownership, which matured as painfully as an evil boil.

In July 1833 George Gower died, very probably the largest canal and railway proprietor in England, and in terms of acreage the largest landowner in Britain.[2] Owing mainly to his wealth, but also to his ability and services rendered, he had raised the Sutherland title to a dukedom, granted just before his death. The Bridgewater will had left Gower's second son the income from the Trust on condition that he changed his name, and Lord Francis Egerton, as he then became, was a man in whom much of the genius of two able families had blended – the charm and ability of his Gower grandfather, and the Egerton sense of service, perseverance and business acumen. As an added bonus he was endowed with the good looks that his poor old great-uncle had so singularly lacked.

By 1833 he had decided to live as the duke had generally done, for a part of each year at Worsley, in the firm and publicly stated conviction that his 'possessions imposed duties upon him as binding as his rights'.[3] Clearly it was impossible for the new heir and his charming wife to cohabit with the wayworn and crotchety Bradshaw, who had complicated matters by buying a good deal of land of his own around Worsley. The Sutherlands' agent, James Loch, then visited the canal offices to look through the accounts, and make 'some representations on the matter'.[4] The eviction seems to have been handled somewhat harshly, and there was a ruthless element in Loch, though Lord Francis must also have been responsible, but Bradshaw agreed to retire on full salary, and sell his Worsley land at a comfortable valuation. Yet he still had one arrow left in his quiver – the right under the old duke's will to appoint his own successor – and he appointed *not* his own son, who, he probably feared, would sell out to the railway interests of the Leveson-Gowers, but James Sothern, one of his own staff. Bradshaw was endowed with a kind of perverse honesty, and his loyalty lay still with the old duke rather than with the new and still untried heir.

His son, Captain James Bradshaw RN, had supported and worked for his dedicated but difficult father for some twenty years, with the sole consolation that he would probably succeed to the most enviable and independent stewardship in the land. The strain proved too great on a mind already worried and depressed, and he committed suicide by cutting his own throat, with a razor, in the Brick Hall.

Meanwhile an inherent decency had prevented Lord Francis Egerton from so much as asking his father about the terms of the old duke's will, and he was staggered to learn that Bradshaw was empowered to appoint Sothern. He wanted to move into the Brick Hall as a temporary seat, but Sothern refused even to allow his furniture into the place.[5] A solicitor called Varey was sent to spy on Sothern, and a host of twelve charges were trumped up against him, in a case in which the judge early intervened, sending both parties packing with an injunction to settle the matter privately. This was done with £40,000 compensation to the agent, who retired to Liverpool and also bought an estate in Yorkshire. James Loch, the trusted steward of the long and losing battle to make the county of Sutherland into a viable economic unit, was appointed Bridgewater Superintendent, and by 1837 the stage was set for the reign of another Francis Egerton, this time aided by Harriet, his wife. She has been described as the 'classical Lady Bountiful, a woman of strong Christian conviction, who won the confidence of the miners and their wives in a truly remarkable fashion'.[6] Since neither of them wished to live in the Brick Hall, now associated with the suicide, all but the stables of the duke's old house was pulled down, and Edward Blore was commissioned to build the Gothick New Hall just across the road which he undertook in 1846.

When Lord Francis Egerton settled at Worsley he was intrigued to learn the extent to which the old duke – always something of a joke to his own family – was admired locally, and the degree to which he had impressed his personality on the surrounding countryside. This popularity lay, as he soon realised, more with the poorer people than with the squires and Manchester and Liverpool merchants, whose opposition had sometimes been so adamant.

Concerned that there was no monument to his illustrious benefactor, Egerton had approached the Lord Mayor of Manchester, and in 1835 John McVicar replied, 'A very slight acquaintance with our smoky town . . . will acquaint you with how very few eligible situations we have for a statue. The *best* site I should think is in the Pond in front of the Infirmary – a bronze statue there would be very well placed.'[7] All the more so, perhaps, as the duke had been president of that institution. A committee was formed to select an appropriate site, and decided that Thomas Campbell should be invited to visit Manchester as soon as possible. Eight years dragged by while Campbell completed other commissions, but by July 1843 he was writing to say that he hoped to cast the very much more than life-size figure – 13 ft 3 in. without any plinth – within a few days, and that he would like it to be placed somewhere over the Bridgewater canal. 'My whole study in this case,' he wrote, 'was to create an effective monument to a nobleman whose whole life was devoted to science.'[8]

Lord Francis was annoyed, not merely by the failure to get the job done on time, but by the more than doubled cost of the project – due mainly to the sculptor's insistence on such an enormous figure. He felt that family piety could scarcely expect more than £2,000, as originally agreed, and this was forfeit when he refused to pay more, so the figure was never cast.

Worsley, under Bradshaw's later reign, had sunk into 'religious and educational destitution'.[9] Under the new squire all this was rapidly changed. More schools were started, the employment of women and girls underground in the mines was abolished, and in Parliament Lord Francis supported Shaftesbury's Act of 1842, which made this type of labour illegal. Night schools, mining safety classes, libraries, reading rooms with paid organisers, and a host of other benevolent activities soon turned Worsley into a model village of the industrial revolution. When Lady Harriet died in 1866, 'possessed of all the virtues that adorn a woman',[10] a monument was generously subscribed to by the local people similar to the one raised to her husband on his death in 1859.

Francis had been granted the earldom of Ellesmere in 1846, and throughout the long Victorian era the tradition of benevolent estate management was maintained, first by his son George, and then by Algernon Egerton, who was Bridgewater Superintendent from 1855 to 1891. It was a good record, smelling sweet in the dust of history alongside the intransigently grim archives of the 'classic slum', only a few miles away in parts of Manchester, Eccles and Salford. The acid test of the Chartist agitations engendered a remarkable document addressed to Lord Francis in which the miners emphasised their willingness to 'defend your honour and all in connexion', and expressing their 'gratitude and love towards you'[11] – co-operation between labour and management scarcely equalled even during the duke's day, and never to be seen again.

Meanwhile, though the profits of the Bridgewater did not greatly increase, it held its own against the railways, and by 1860 was carrying nearly three times as much freight as it had transported in 1833, when the duke's younger nephew had become the beneficiary. Loch bought the Mersey and Irwell Navigation in 1844 for £402,000 and the Trust was broken in 1872, when the canal and the old waterway were sold to the Bridgewater Navigation Company, which in turn sold out to the Manchester Ship Canal Company, its present owners, in 1887, for £1,710,000. This company's Act laid down that the Bridgewater should be kept thoroughly repaired, dredged, in good working order, and open for navigation to all who wished to use it.

During the Kaiser's war the Bridgewater Canal continued to do yeoman service, and by 1923 the Egertons had sold virtually all their local estate interests to the present owners, a property concern called the Bridgewater Estate Company. By 1940 the old canal was still carrying nearly a million tons, and remained viable financially into the early 1960s, but the closure of the Lancashire coal mines, combined with the encouragement of road and rail transport, resulted in a small deficit by 1970. Runcorn docks continue to prosper, though now detached from the canal, and there has been a remarkable resurgence of life as pleasure boating and angling have taken over where the 'dukers' once plied their trade.

The towns of Manchester, Salford and Liverpool never saw fit to honour with a memorial the technocrat who had contributed so much to the solution of the fuel crisis, but at Ashridge the curious visitor may climb a column designed by F. H. Egerton, and gaze down across the Grand Union Canal which links London with the Midlands. Brindley has a plaque at his birthplace, while in Worsley there are two others commemorating the achievement of these early canal builders and marking a bicentenary of the first sale of coal across Barton aqueduct. The small peninsula running out from the wall of the Delph is now cleared again and terraced, with one of the starvationers from the sough stationed nearby. Worsley New Hall has vanished, like the Brick one in which Captain James Bradshaw died so terribly, but the Old Hall is still carefully preserved, and on Worsley Green the base of the tall chimney at the duke's canal-side works was converted to a fountain. On its steps was engraved, in somewhat latter-day Latin, some verses of which the following lines seek to capture something of the original spirit:

Bred of Cyclopean skill, and iron master's will
 I soared, a columned chimney, smoking high.
That Duke had been my author, whose broad bridge,
 Spanning the floods beneath, his name we can imply.

Consider, stranger, how this gravely peaceful place
 Roared with Chalybean stridence and unquiet.
Only my bubbling waters now can trace
 From underneath this font, a rise and fall of Grace.

Alas, we both, who overlooked these lands
 So loftily, now shrink to this sad fate!

So – lest our memory altogether die
Behold! Some part of me will testify
That what remains bears witness to what was.

The Cyclops – in Homer a race of one-eyed monstrous giants – were changed into craftsmen by an evolving mythology, bound apprentice to Vulcan on Etna, and forgers of Zeus's thunderbolts. The Chalybes were famed ironworkers of antiquity, living on the south-east coast of the Black Sea – much iron was cast in the duke's furnaces, and the sough waters are chalybeate.

The pipes which once conveyed this water to the fountain are still furred up, but the time may yet come when it will 'leap from hidden springs and fall again' in honour of 'the father of inland navigation', who turned the streams and rivers of England to such an admirable service.

Notes to this chapter are on p. 197

Mr John Antrobus, Aston
To His Grace the Duke of Bridgewater for Freight &c. Dr

Date 1793	Species of the Goods Freighted	Weight T C Q lb	Toll S D	Freight L S D	Paid on Rec't of Goods L S D	Cartage L S D	L S D	Total Amo't of Freight & Charges L S D
June 4	Atalker 14 Lds Meal	1 10 0 0	5/8	0 8 6		0.2.1		0 10 7
— 6	Do 16 Do	1 14 1 4	"	0 9 8		0.2.5		0 12 1
— "	Do 7 Lds Flour	0 7 5 0 0	.	0 4 3		0.1.1		0 5 4
							£	1 8 0
	Per Contra Cr.							
June 11	By Cash	£1:8:0						

19 Bridgewater Canal freight bill found at Bartington, Cheshire

Appendices

Appendix A. Cal. 10329. In SEM. Some account of the Adits at Worsley with their Branches, 17 June 1772.

1. The Navigable Adit makes its Enterance [*sic*] from the Canal at the stone quarry at Worsley Mill and runs Northward into the local Mines, begun in the Year 1759 the whole length of which is 2520 yards.
2. Another larger from the same Quarry was begun in May 1771 to communicate with the former at the distance of about 500 yards and will form a Scalenous Triangle between the Quarry and the two Adits.
3. A Branch from the Adit that runs West to near Ellinbrook begun about 1761 and sets off near 777 yards North of the Quarry. The length of which branch is 1805 yards.
4. Another Branch Westward that sets off 1320 yards more North and begun March 1770 called the Bin Coal Level. The Length of which is 488 yards.
5. Another Branch Westward that sets off about 205 yards more North than the last Branch and is called the Crumback Coal Level, begun in 1771, the length of which is 210 yards.
6. Another Branch Westward that sets off near 108 yards more North than the last Branch and is called Brassey Coal Level, begun in Septr. 1771. The length of which is 66 yards.
7. Another Branch Westward that sets off about 118 yards more North than the last Branch and is called the 8 foot Coal Level, by May 1772.
8. The Old Sough made many years ago, for Draining the Coals under Walkden Moor was about 1766 begun to be Widen'd and extended across that Moor from South to North, which Line is near a Measured Mile and Quarter in Length, of which 210 yards as near as can be Computed remains undone; about 160 thereof are on the South side the Moor and when perfected down the Great Road on that side the Moor it is there proposed to Rest, and to continue the Adit to the same Spot, which will be there about 60 yards below the surface and about 28 below this Sough.

The Coals will be conveyed from the Upper to the lower Level and Navigated into the Canal at the Quarry before mentioned in like manner as they are from the Branches.

I have read this account of the Adits Branches &c & I think the same is right (signed) John Gilbert ?

[This signature is almost illegible, but *seems* to be J. Gilbert senior's]

Appendix B. SEM, Sketch plan of the Bridgewater Canal by Brindley

Legend in key: 'note the length of the old River from manchester Warington is about 30 Miles and the fall of water from manchester to Warington is 52 poles 5 inches . . .' Beside this – 'note the Canal is marked thus' (double line) 'and the old River is in Black' (single line). 'The length of Duke of Bridgewater Canel from Worsley to Manchester and from manchester to Runcorn is measur'd 30 Statute miles . . .' A further note on bluish paper attached to the map gives what may be a MS sale number, and states: 'Original Plan of Duke of Bridgewater's Canal from Worsley, Manchester to Runcorn by Brindley, the Engineer, given by Duke to Capt. Dewhurst . . . is curious and of great value.'

Appendix C. SEM, Death mask

The box at Mertoun contains the legend: 'Death mask and Casts of the Ears of Francis, 3rd Duke of Bridgewater (Removed from Bridgewater House, St James' SW on Dec 20th 1919)'. On the lid: 'Cast of Duke of Bridgewater, March 10th 1803'.

Appendix D. SEM, Note by Robert Lansdale, 13 January 1844.

A useful brief summary of progress on the canal's construction. 'The first part of the Canal from Worsley towards Manchester would be opened as far as Barton in 1760–1. Barton Bridge (the aqueduct) would take 2 years in erecting = whilst the Bridge was erecting, coal boats came up to Barton and the coals were let down by a Gig Crane, in twig baskets to vessels in the Irwell (& so to Manchester) There was a feast of roast beef &c to the Men on completion of the Arches & it being about the time of the Coronation of G. 3rd they said they had crowned the Arches: in 1763 coal boats came up to Cornbrook Bridge, discharged & sold there, & in 1764 the same year as the lands in Hulme were purchased from George Lloyd Esq. they arrived at Castlefield with their cargoes. Worsley Coal was sold at Preston Brook in 1771.

The Runcorn Locks were finished at Lady Day 1776 and flats first sailed through from Liv. to Manch.

The adit was begun in 1759 at Worsley Mills

It may be asked how did flats built at Worsley get into the Irwell when there was no road at Runcorn – at Cornbrook there was a Sluice or Gut, cut into the river there, and they were let down by it into the River.

Once, before Mr Gilbert died, his Grace came into Lancashire, but his barge had not arrived (by mistake) at Broadheath to meet him, so instead of taking Post Horses forward, he ordered his Carriage to be put into an empty coal lighter, got into it, & was hauled up by his mules to Worsley.'

Appendix E. Extracts from Strachan Holme's Supplementary Notes, etc. to SH, Sec. I, 2 ff.

'The difficulty in understanding why Francis, late Duke of Bridgewater, has no adequate or complete biography arises from a variety of causes:

1. That he was far too great a man for some of those who essayed the task.
2. From his own reticence or indifference to other people's opinions as concerned himself.
3. The almost universal lack of sympathy and appreciation in men of his own class and even of the earliest beneficiaries under his Will, to his enthusiasm for canal-making and the development of trading facilities in England.
4. Their ignorance of his motives and intentions – even of his testamentary dispositions.

The rather egregious Smiles . . . made rather too great a figure of Brindley, as the extracts from Eliza Meteyard's 'Life of Josiah Wedgwood', The Rev Francis Henry Egerton's writings, and other sources indicate. Smiles made enquiries at the Bridgewater Trust Offices, Manchester, for information about the canal making, Brindley's remuneration, etc., and was informed that there was no information there. There was, however, and it was turned out some years later.

In his own times those who knew him would think it quite unnecessary to write any life of him, as on the lips and in the hearts of the people he was familiar and greatly respected as a great bringer of employment and all the comforts and well being that accompany it.'

Appendix F. The Canal debt, taken from EB 1461, General State of his Grace the Duke of Bridgewater's Navigation, &c

From Midsmr.	1759–Xmas 1760	£8,568	10	5
to	Xmas 1762	£27,701	7	3
to	Sept 1764	£45,299	15	8
to	Nov 1765	£60,879	2	5
to	Oct 1766	£73,645	11	9
to	Mar 1768	£92,535	3	1
to	26 May 1769	£108,159	11	9
to	29 Sept 1769	£112,322	1	3
to	20 Oct 1770	£133,219	15	2

	Date	£	s	d	
to	6 Oct 1771	£153,448	13	8	
to	1 Jan 1772	£162,605	12	9	
to	9 Jan 1773	£188,529	2	11	
to	25 Dec 1773	£210,115	13	1	
to	7 Jan 1775	£229,212	19	4	
to	13 Jan 1776	£246,087	0	7	
to	10 Jan 1777	£263,724	11	7	
to	10 Jan 1778	£280,031	13	0	
to	Jan 1779	£294,279	19	1	
to	12 Jan 1780	£311,207	18	2	
to	9 Jan 1781	£319,927	2	0	
to	11 Jan 1782	£327,471	8	9	
to	9 Jan 1783	£339,875	15	8	
to	8 Jan 1784	£344,511	0	5	
to	11 Jan 1785	£345,173	3	5	
to	14 Jan 1786	£346,805	18	3	
to	10 Jan 1787	£346,371	10	8	*Debt decreased this year 434–7–7*
Highest point of the debt		*£346,805*	*18*	*3*	
to	19 Jan 1788	£342,874	12	7	Debt decreased this year £3,496 18 1
to	19 Jan 1789	£340,937	9	6	Debt decreased this year £3,931 5 8
to	13 Jan 1790	£337,401	0	6	Debt decreased this year £3,536 9 0
to	10 Jan 1791	£329,175	6	2	Debt decreased this year £8,225 14 4
to	1792	£324,745	14	3	Debt decreased this year £4,429 11 11
to	1793	£311,424	6	3	Debt decreased this year £13,321 8 0
to	1794	£321,229	12	1	*Increase* in debt £9,805 5 10
to	1795	£325,099	15	11	*Increase* in debt £3,870 3 10
to	1796	£321,030	1	6	*Decrease* in debt £4,069 14 5
to	1797	£303,701	13	1	*Decrease* in debt £17,328 8 5
to	1798	£290,782	16	7	Decrease in debt £12,918 16 6
to	1799	£283,124	17	6	Decrease in debt £7,657 19 1
to	Jan 1800	£271,143	14	7	Decrease in debt £11,981 2 11
to	1801	£253,732	8	6	Decrease in debt £17,411 6 1

to	Jan 1802	£226,341 19 4	Decrease in debt £27,390 9 2
to	1803	£162,397 14 1	Decrease in debt £63,944 5 3
	1804	£104,565 3 8	Decrease in debt £57,832 10 5
	1805	£47,530 17 9	Decrease in debt £57,034 5 11
	1806	Profits below	
Difference being an Increase of Profits Jan 1806 over and above paying the whole Principal that the Navigation cost, with the Acct. of the Interest thereof added thereto Annually at 5 per cent.			£20,935 4 0

At the year 1804 is the following entry in red ink: 'This account was made up to the Trustees of the late Duke of Bridgewater: His Grace having died on 8 March 1803'.

Appendix G

The Tatton miniature of the Duke is a fine example of its kind – a water colour on a thin ivory background. The artist is now known to have been Richard Crosse, and the portrait is dated *c.* 1790. The mount is probably of gold, studded with emeralds in a cross, on a field of diamonds.

Behind glass on the reverse is a lock of the duke's fair hair, held by wire and three pearls, with two wisps braided into an ear of corn and a flower by animal gut, on a silk background.

Although carefully verified at the time, the authenticity of the frontispiece of the first edition of *The Canal Duke* is now in some doubt.

I am indebted to George Clark, House Manager of Tatton Park, Marion Kershaw and Ahmed Youssef for allowing me to study this artefact.

Notes

Chapter 1. The homeless duke

1 British Waterways Board news release, 26 November 1973, q. Sir F. Price, chairman.
2 HPT, R. Wood to Duke of Bedford, 18 May 1754.
3 For his life see *DNB* with bibliography; *Chalmer's Biog. Dict.*, London 1814; SEM generally, and a reassessment in SH, *Family Chronicles*.
4 SEM, *Notes of proof for Lord Ellesmere* – Historical and General, *c.* 1919. Among properties the Egertons were obliged, as Royalists, to sell, was 'the Manor of Bridgewater in Somerset from which it is presumed the title was taken'. The second earl was also involved in the misfortunes of his son-in-law, Sir William Courten, a Merchant Adventurer.
5 SEM Cal. 10260. I am grateful to Peter Walne, the County of Hertford Archivist, for identifying 'Markyate' with Market Street, Herts.
6 Granville Leveson-Gower, Viscount Trentham, later Marquis of Stafford, m. Lady Louisa Egerton 28 March 1748.
7 CRO Herts, AH 1464, and R. Clutterbuck, *History and Antiquities of The County of Hertford*.
8 Beatrix. Details from archives at Tatton Park in Cheshire, where her portrait hangs.
9 Tatton Park Guide, National Trust, 1972, edition. W. H. Chaloner. 'The Egertons in Italy and the Netherlands, 1729–1734', *Bulletin of the John Rylands Library*, vol. 32 (1949–50), 157.
10 SEM, Chancery case, 31 October 1748 (Cal. 10260), Draft of Complaint of Francis, Duke of Bridgewater, by the Duke of Bedford, his next friend. . . .
11 *Ibid.*
12 Quoted from the Tatton papers, now with CRO Cheshire. I am grateful for the information from Mr Patrick Strong, Keeper of Eton College Library and records, that Dr Purt was a private tutor of the kind attending most noblemen there, up to the nineteenth century. Francis may have been at Eton in 1749–51.
13 HPT, Duke of Bedford to Sir R. Lyttelton, London, 14 April 1751.

14 SEM, Chancery case, *op. cit.*, throws much light on the duke's early character and determination.
15 7 Geo. I, C 15. *CNW*, 16.
16 *CNW*, 17.
17 26 Geo. II, cap. 63; V. I. Tomlinson, 'Salford activities connected with the Bridgewater Navigation', *TLC*, vol. 66 (1956), 56; *CNW*, 17.

Chapter 2. The grand tour

1 W. H. Chaloner, 'The Canal Duke', *History Today* (October 1951).
2 SH, Sec. 1, 7.
3 SEM, Wood to Gower, Paris, 19 March 1753.
4 *Ibid.*
5 SEM, Wood to Gower, Lyons, 25 May 1753.
6 HPT, Bridgewater to Bedford, 17 May 1753.
7 SEM, Wood to Gower, Lyons, 30 May 1753.
8 SEM, Wood to Gower, Lyons, 7 June 1753.
9 *Ibid.*
10 *Ibid.*
11 HPT, Wood to Bedford, 16 November 1753.
12 The Canal du Midi is admirably chronicled in L. T. C. Rolt, *From Sea to Sea* (1973).
13 W. A. McCutcheon, *The Canals of the North of Ireland* (Newton Abbot, 1965), 18.
14 As Professor J. R. Harris has shown, England tended, in some other technical matters, to be ahead.
15 HPT, Wood to Bedford, Lyons, 18 May 1754.
16 HPT, Bridgewater to Bedford, 24 September 1754.
17 HPT, Wood to Bedford, Nice, 10 October 1754.
18 HPT, Wood to Bedford, Rome, 9 January 1755.
19 HPT, Wood to Bedford, Rome, 24 May 1755.
20 HPT, Wood to Bedford, Rome, 8 September 1755.

Chapter 3. Of soughs and ale

1 W. H. Chaloner, 'The Canal Duke', *History Today* (October 1951), 66; SEM, *Town and Country Magazine* (1775); SH, Sec. 7, 9.
2 J. H. Plumb, *England in the Eighteenth Century* (1961 edn.), 11 and 78.
3 SH, Sec. 7, 3.
4 SEM, J. Benson to J. Loch, 25 December 1843.
5 SEM, Extracts from *The Farington Diary* by Joseph Farington R.A. (*c.* 1798–1803), mentions Cullen Arabian as bred in 1757; Astridge may be really Ashridge.
6 Bernard Falk, *The Bridgewater Millions* (1942), 90.
7 SEM Cal. 10263/4, 1764, probably written by the duke.
8 Chaloner, 'The Canal Duke', *op. cit.*

9 SS, 312.
10 SEM – an undated copy only, *c.* 1757, not giving exactly the original wording. EM, p. 6, gives the date of the broken engagement as November 1758.
11 SEM, copy of Mrs Wood's letter, *op. cit.*
12 *Ibid.*
13 *Ibid.*
14 *Ibid.*
15 SH Sec. 7, 6. Horace Walpole, fourth Earl of Orford, 1719–97, whose memoirs offer a valuable insight into his times, *cf.* W. S. Lewis, ed., *The Yale Edition of Horace Walpole's Correspondence* (1937 onwards).
16 SH, Sec. 7, 5.
17 SEM, *An Abstract of Deeds Evidences and Writings* . . . of the duke's various estates completed in 1761, by his orders, to ascertain his resources, 433 pp.; CRO Herts., Tyler's accounts for 1754–56.
18 CRO Herts. This amounted to £400 p.a.
19 CRO Herts, AH 1873–1884.
20 SEM, *An Abstract of Deeds*, *op. cit.*, includes a lease, p. 351, from Pierce Sharkie granting the duke the right to drill a sough and extract coal and minerals. 28 December 1759.
21 The evidence for this figure is tenuous, but it seems a possible sum.
22 Will of the Duke of Bridgewater, British Museum (printed 1836).
23 Frank Mullineux, *The Duke of Bridgewater's Canal* (Eccles, 1959), 6.
24 BES, Massey's Account Books, 1721–45, with some years missing (exact dates were not always given in these books).
25 *Ibid.* Massey's salary remained at £50 p.a. throughout.
26 BES, 1725.
27 BES, 1735.
28 BES, 13 August 1726.
29 EM, 85; Frank Mullineux, in *The Great Human Exploit* (1973), 48 – a record of coal at Worsley in 1376.
30 BES, Massey's accounts, 13 September 1735.
31 BES, *op. cit.* 14 January 1726.
32 BES, *op. cit.*, 1726.
33 BES, *op. cit.*, 1731.

Chapter 4. The canal idea

1 T. S. Willan, *River Navigation In England 1660–1750* (2nd edn., 1964), 2 ff, establishes the considerable quantities of coal and other goods carried on early navigations; C. Hadfield, *British Canals* (4th edn., 1969), chapter 1. W. T. Jackman, *The Development of Transportation in Modern England* (2nd edn., 1962), 356.
2 T. C. Barker, 'The Sankey Navigation: the first Lancashire canal', in TLC, vol. 100. The precedence of the Sankey as a distinctive canal was opposed by Johan Weale in *Quarterly Papers on Engineering*, vol. 1, n.d. (in MLH), and by Lord Francis Egerton in his essays contributed to *The Quarterly Review*, 1858, quoting

Hughes's opinion that the Bridgewater ran 'not in the bed of a stream'. SH, Sec. 7, 7. Lateral, summit-level canals, and navigations are well defined in L. T. C. Rolt's *From Sea to Sea*, 3 ff.

3 *CNW*, 15–16.
4 7 Geo. 1, cap. 15.
5 BES, Massey's accounts, *op. cit.*, 1725.
6 *CNW*, 17.
7 SEM Cal. 10319, The Proposals of The Mersey & Irwell, or Naviagors [*sic*], n.d.
8 *Ibid.*
9 *Ibid.*
10 *Ibid.* Each basket weighed 182 lb or more, loaded.
11 10 Geo. 2, cap. 9.
12 Tomlinson, 'Salford activities', *op. cit.*, 54.
13 HLRO, 32 & 33 Geo. 2; JHC, vol. 28, 1757–61, p. 321, 25 November 1758.
14 *CNW*, 19.
15 *CNW*, 21.
16 EM, 6.
17 JHC, vol. 28, *op. cit.*
18 *Bridgewater Canal Guide*, 1.
19 H. Malet, 'Brindley and canals', *History Today*, vol. 23, No. 4, April 1973, 266 ff.
20 SEM, R. Lansdale to J. Loch, Worsley, 21 December 1843.
21 EM points out that Brindley also worked at Tatton. Ordsall was inherited by S. Egerton in 1758 from his uncle, Samuel Hill – *Ordsall Museum Guide*, 3.
22 *DNB*, Gower.
23 R. M. Larking, *The Canal Pioneers* (1967, published privately), 31. I am grateful to Mr W. Howard Williams for considereable additional help with details of the Gilbert family.
24 *DNB*, Gilbert.
25 Larking, *op. cit.*, 33.
26 *Ibid.*
27 *DNB*.
28 Gilbert's Act, 22 Geo. III, cap. 83; some of his publications are in the British Museum Library. W. E. Tate, *The Parish Chest* (Cambridge, 1946), 227. His Act sought to humanise the Poor Law.
29 W. E. Tate, *op. cit.*, 229.
30 I am grateful to W. Howard Williams for this correction.
31 W. H. Chaloner, *People and Industries* (1963), 34.
32 CRO Bedford. By 1759 Thomas Gilbert was witnessing and administering Ashridge estate deeds at Ashridge, but not in 1751.
33 SEM, R. Lansdale to J. Loch, Worsley, 21 December 1843.
34 SEM Lansdale, *op. cit.*
35 SEM, R. Statham to Mr Taylor, 12 February 1790, for Gilbert's humour. T. Gilbert to Francis Gildart, 28 January 1777, ref. bonds mislaid.
36 A. Rees, B.D., ed., *The Cyclopaedia, or Universal Dictionary* (1819), Gil; Larking, *op. cit.*, 21 ff.
37 EM, 6.

38 TLC, T. C. Barker, 'The Sankey Navigation', *op. cit.*, 137.
39 H. Malet, 'Brindley and canals', *op. cit.*, 268.
40 A. Rees, *op. cit. The Cyclopaedia*, Gil and Canal; SEM, Robert Lansdale to J. Loch, 21 December 1843.
41 SEM, F. E. Egerton, *The First & Second Part of a Letter . . . Upon Inland Navigation.*
42 J. R. Harris, 'Liverpool canal controversies, 1769–72', *Journal of Transport History*, vol. 2 (1955–56), 171.

Chapter 5. The first Act

1 A. Rees, *Cyclopaedia, op. cit.*, Gil.
2 John Rylands Library, Manchester. Appeal To the Gentlemen and Traders of Manchester and Salford (n.d.; October 1758).
3 HLRO, original Act, 32 & 33 Geo. II. JHC, vol. 28, 1757–61. I am grateful to Mr Maurice Bond, Clerk of the Records, House of Lords, and his staff for their kind guidance.
4 *Ibid.*
5 *Ibid.*
6 I am grateful to Mr Frank Sharman for pointing out that navigations and even enclosures had involved some compulsory purchase; but the mild deception in planning the Sankey Navigation shows that Parliament was not sympathetic, even to lateral canals.
7 HLRO, JHC, vol. 28, *op. cit.*
8 *Ibid.*
9 I am grateful to Mr Arthur Lucas of Woolmers, Herts, for guidance on 'levelling'.
10 JHC, vol. 28, *op. cit.*
11 SH thought that ten locks would be required, which explains the reluctance of the Mersey & Irwell to undertake the work, since their navigation was not paying well.
12 JHC, vol. 28, *op. cit.*
13 *Ibid.*
14 SEM, Agreement . . . Proprietors of the Irwell, & Duke of Bridgewater's Agent, 14 June 1759.
15 *Ibid.*
16 HLRO, JHC, vol. 28, *op. cit.*, 5 February 1759.
17 *CNW*, 21.
18 HLRO, JHL, vol. 29, 433.
19 *Ibid.*
20 HLRO, Committee Book HL, 7 March 1759.
21 TLC, Tomlinson, 'Salford activities', 65.
22 CRO Herts.
23 TLC, Tomlinson, 'Salford activities', 63.
24 Manchester Central Library, *The Constables' Accounts of Manchester, 1612–1776* (Manchester, 1891). I am grateful to Christopher Makepeace and his staff for their help at the Central Library, Local History Department.

25 W. T. Jackman, *The Development of Modern Transportation*, 3rd edn., (1966), 360.
26 TLC, Tomlinson, 'Salford activities', 63.
27 SEM, *An Abstract of Deeds, Evidences* &c. . . ., *op. cit.*
28 SEM, J. Fenton to J. Loch, Stoke Lodge, 19 December 1843.
29 Jackman, *op. cit.*, 188–9.
30 CRO Northants., in which some early pages of account books are left blank. Also, Rees, *op. cit.*; SEM, Note by Robert Lansdale, 13 January 1844.

Chapter 6. The canal triumvirate

1 SEM, R. Lansdale to J. Loch, Worsley, 21 December 1843.
2 H. W. Dickinson, *Matthew Boulton* (Cambridge, 1937).
3 J. P. Wadsworth and J. de Lacy Mann, *The Cotton Trade and Industrial Lancashire* (Manchester, 1931), 170. J. Aikin, *England Delineated* (1788), 76 ff.
4 J. Aikin, *op. cit.*, 76 ff.
5 *Ibid.*
6 SEM, R. Lansdale to J. Loch, 21 December 1843.
7 James Brindley's diary, Birmingham Public Library, from 19 September 1756 to 27 June 1758.
8 Rev. J. Fenton to J. Loch, Stoke Lodge, 19 December 1843.
9 Examples include SEM, Cal. 10281 and 10283.
10 I am grateful to Professor W. H. Chaloner and Mr M. Gleeson for a copy of Brindley's expenses for survey work, 13 November 1761; this document is quoted in SS also.
11 CRO Northants, EB 1461, General State. . . .
12 Rev. F. H. Egerton, *First & Second Part of A Letter . . . Upon Inland Navigation* (Paris, n.d.), 66.
13 H. Bode, *James Brindley* (Leek, 1972), 3.
14 C. T. G. Boucher, *James Brindley, Engineer* (Norwich, 1968), 65. Hugh Malet, 'Brindley and canals', *History Today* (April 1973), 267.
15 A. G. Banks and R. B. Schofield, *Brindley at Wet Earth Colliery* (Newton Abbot, 1968), 15.
16 Brindley's diary, 21 Aporil 1758.
17 C. Hadfield, *The Canals of the East Midlands*, 2nd edn. (1969), 18–19.
18 J. H. Campbell, *Derbyshire Miscellany*, February–June 1956–9, MLH.
19 This map of 'The Duke Bridgewater Canel', as Brindley captioned it, was given by the duke to a Captain Dewhurst. It is now in the Collection of the sixth Duke of Sutherland.
20 Boucher, *op. cit.*, 24–5.
21 SEM, Evidence for the third Act, 28 January 1762, in which Brindley claimed to have been engaged on the work since it began, meaning most probably the survey of the Trent and Mersey.
22 CRO Northants, EB 1459, Day Book of the Duke of Bridgewater. (The title page of this work seems to have been lost, possibly in rebinding.)
23 W. H. Chaloner, 'James Brindley and his remuneration as a canal engineer', TLC, vols. 75 and 76 (1965–66), 226.

24 CRO Northants., EB 1459 and EB 1461.
25 SEM, Draft petition for the second Act.
26 HLRO, JHC, 1757–61.
27 HLRO, JHC, 31 January 1760.
28 Tomlinson, 'Salford activities', *op. cit.*, discusses motives on p. 71.
29 Tomlinson, *op. cit.*, 73.
30 L. T. C. Rolt, in *The Inland Waterways of England*, 5th edn. (1970), like other writers, attributes more to Brindley than the latest evidence warrants.
31 HLRO, 33 Geo. II, cap. 2, passed 24 March 1760.
32 CRO Northants., EB 1459, Day Book, *op. cit.*
33 *Ibid.*
34 T. Lowndes, *The History of Inland Navigations, Particularly . . . Duke of Bridgewater in Lancashire & Cheshire*, 2nd edn. (1769), 16.
35 A later *Commissioners' Minute Book* survives – CRO Northants, EB 1469.
36 CRO Northants., Day Book, *op. cit.*
37 SEM, Note by R. Lansdale, 13 January 1844.
38 CRO Northants., EB 1459, Day Book, *op. cit.*
39 SEM, Note by R. Lansdale, 13 January 1844.
40 SS, 390 (1862 edn.).

Chapter 7. The third Act

1 CRO Northants; CRO Herts, where there are records of these bonds.
2 SH, Sec. 7, 15.
3 CRO Northants, *General State of His Grace The Duke of Bridgewater's Navigation, Colliery, Lime and Farm Concerns in Lancashire & Cheshire, Midsummer 1759*, EB 1461.
4 CRO Herts, though, offers one or two clues in the A. H. collection.
5 *DNB*, Fowler.
6 I am grateful to Mrs N. Burns for her letter of 21 June 1962, mentioning the recommendation of the duke in Lambeth Palace Library (W4/20/32F certificates of contracts, 1733–1802), stating wryly that Fowler 'never did have any advantage' from being in his service.
7 Rev F. H. Egerton, *First & Second Part of A Letter*, *op. cit.* (Paris, n.d.), 62.
8 SEM, Robert Lansdale, Note of 13 January 1844.
9 T. Lowndes, *The History of Inland Navigations*, *op. cit.*
10 From James Ogden, *A Description of Manchester* (1783), Chetham's Library, Phelps collection. *CNW*, 26. Charles Cotton (1630–87) wrote *Wonders of the Peak*, 1681.
11 SEM, Queries concerning the Old & New intended Navigation from Manchester to Liverpool, n.d.
12 *Ibid.*
13 SEM, Queries . . . the answers opposed thereto . . ., n.d.
14 SEM, Reasons of the Proprietors against the Bill . . ., n.d.
15 Reasons, *op. cit.*
16 HLRO, JHC, 14 November 1761.
17 SS, 231.

18 HLRO, JHC, 14 November 1761.
19 SEM, Observations . . ., n.d.
20 *Ibid.*
21 *Ibid.*
22 SEM, Parliamentary evidence for the third Act.
23 *Ibid.*
24 *Ibid.*
25 SEM, R. Lansdale, Note of 13 January 1844.
26 SEM Cal. 10281, 28 March 1797, from Cleveland Court.
27 Huntingdon Library, California, Parliamentary evidence for the third Act.
28 *Ibid.*
29 *Ibid.*
30 *Ibid.*
31 HLRO, JHC, 14 March 1758. I am grateful to Charles Hadfield for guidance on early lock construction.
32 SEM, Evidence for the 3rd Act, 28 January 1762.
33 *Ibid.*
34 Huntingdon, *op. cit.*
35 SEM, Parliamentary evidence, 2 January 1762.
36 Huntingdon Library, *loc. cit.*
37 Malet, 'Brindley and canals', *History Today*, 270.
38 *DNB*, Derby, Earls of. James Stanley-Smith, wrongly entitled Lord Strange, born 1716, was Lord Lieutenant of Lancashire, and died of apoplexy at Bath in 1771. It has been pointed out that this branch of the Stanleys was not really entitled to the barony of Strange – *cf.* Vicary Gibbs, *The Complete Peerage* (1916), 216.

Chapter 8. The underground canals

1 *Gentleman's Magazine*, vol. 36 (1766), 31.
2 Henri Fournel and Isidore Dyèvre, *Memoire sur les canaux souterrains . . . de Worsley* (Paris, 1842), 15. I am grateful to Dr D. E. Owen, Director of the Manchester University Museum, for lending me his copy.
3 F. Mullineux, *The Duke of Bridgewater's Underground Canals at Worsley*, TLC, vol. 71 (1961), 158.
4 EM, 15.
5 F. H. Egerton, *First & Second Part of A Letter . . . Upon Inland Navigation* (Paris, n.d.; *c.* 1810), 38.
6 EM, 3.
7 CRO Northants.
8 *Ibid.*
9 EM, 110.
10 EM, 115–18.
11 J. R. Harris, 'Liverpool canal controversies, 1769–72', *Journal of Transport History*, vol. 2, 159.
12 I am grateful to Mrs N. Burns, whose father was Bridgewater agent at Preston Brook, for this information.

13 *Manchester Mercury*, 28 April 1762.
14 EM, 105; *Report of the Society for Bettering the Condition of the Poor*, n.d., Brit. Mus., vol. 1, 238.
15 EM, 108, quoting Gisborne.
16 SEM, Mr Peters to J. Gilbert, 17 August 1787.
17 EM, 101.
18 EM, 107.
19 E. R. Hassal and J. P. Tricket, 'The Duke of Bridgewater's underground canals', *The Mining Engineer*, vol. 123, No. 37 (October 1963), 53.
20 W. H. Chaloner, 'James Brindley and his remuneration', *op. cit.*
21 Hassal and Tricket, *op. cit.*, 11 (in the same paper read to Manchester Geological and Mining Society, 15 November 1962).
22 *Ibid.*, 56.
23 *Ibid.*, 47.
24 *Ibid.*, 49.
25 *Encyclopedia Britannica,* Drilling.
26 I am grateful to the Duke of Sutherland and Mrs F. A. Willink for information about these clocks. Mr Mullineux tells me that the mechanism of the church clock is later – probably renewed.
27 *Gentleman's Magazine, op. cit.*, 31.
28 Hassal and Tricket, *op. cit.*, 52.
29 Fournel and Dyèvre, *op. cit.*, 22.
30 *Gentleman's Magazine, op. cit.*, 31 ff.
31 T. Lowndes, *The History of Inland Navigations, Particularly Those of the Duke of Bridgewater and the intended one promoted by Earl Gower and other persons of Distinction in Staffordshire, Cheshire and Derbyshire*, 1st edn. (1766).
32 Rees, *Cyclopaedia, op. cit.*
33 CRO Northants, *General State. . . .*
34 EM, 101.
35 Mullineux, *op. cit.*, 156; Fournel and Dyèvre, 15–16. There were coke ovens at the Walkden pit.
36 J. B. Smethurst suggests that these gases may have been alleviated by the canal waters.
37 (Sir) Joseph Banks, 'Journal of a Journey Through Wales & the Midlands, in 1767–8', Cambridge University Library, Additional MS 6294, NP. I am grateful to Alan Jeffery, and to S. R. Broadbridge of North Staffordshire Polytechnic, for details of the Banks Journal.
38 Banks, 'Journal', *op. cit.*
39 Fournel and Dyèvre, *op. cit.*, 34 ff.
40 Fournel and Dyèvre, *op. cit.*, 61.
41 Lewis's *Manchester Directory* (1788), 17.
42 SEM, Petition, 26 May 1772.
43 CRO Northants, General State. . . .
44 J. H. Clapham, *An Economic History of Modern Britain*, vol. 1 (1964 edn.), 78.
45 *Ibid.*

Chapter 9. The cut goes west

1 SEM, various; SEM, R. Lansdale to J. Loch, Worsley, 21 December 1843.
2 SEM Cal. letter 6, Payment to Brindley for work at Offerdene, signed by W. Bill.
3 SEM, R. Lansdale to J. Loch, Worsley, 21 December 1843.
4 A. Burton, *The Canal Builders* (1972), chapter 11, 126 ff.
5 SEM. The Gilbert correspondence contains many letters – some in draft form – relating mostly to the struggle over the duke's dock at Liverpool. Some of these letters are in the duke's hand – most are between the agents, etc.
6 SH, Sec. 7, 14.
7 SEM, R. Lansdale to J. Loch, 21 December 1843.
8 R. Larking, *The Canal Pioneers*, *op. cit.*, 39 ff.
9 SEM, R. Lansdale to J. Loch, 21 December 1843.
10 Anon., *Remains Historical & Literary connected with the Palatine Counties of Lancaster and Cheshire*, vol. 42 (1899), 27, MLH.
11 (Sir) Joseph Banks, 'Journal of a journey', *op. cit.*, 107.
12 Being denied access to documents, Smiles reached a somewhat mistaken assessment of Brindley's character.
13 'This,' the duke used to say of Brindley's stretch over the Bollin and Mersey meadows, 'is the Worst Part of the Whole Navigation.'
14 The duke was running so short of money in 1762–63 that he was sometimes finding it difficult to pay his staff. There are entries for payment in arrears.
15 SEM Cal., letter 6.
16 CRO Northants, EB 1461.
17 A. Burton, *The Canal Builders*, *op. cit.*, 219.
18 EM, 18.
19 SEM, R. Lansdale, Note of 13 January 1844.
20 *Ibid.*
21 In Malet, 'Brindley and canals', *History Today* (April 1973), the illustration of Barton Aqueduct includes the gig crane, though rickety.
22 Herbert Clegg, 'The Duke of Bridgewater's canal works in Manchester', TLC, vol. 65 (1955), 97.
23 James Sharp, *Extracts from Mr Young's Six Months' Tour Through the North of England* (1774), MLH; Clegg, *op. cit.*, 98.
24 (Sir) Joseph Banks, 'Journal', *op. cit.*, 106. V. I. Tomlinson, 'Early warehouses on Manchester waterways', TLC, vol. 71 (1961). The demolition of the Grocer's Warehouse in 1960 enabled a valuable study of the industrial archaeology of this site. The sough and some machinery was still in place.
25 F. H. Egerton, *First & Second Part of A Letter*, *op. cit.*, 40.
26 CRO Northants; *CNW*, p. 29, argues that it was the duke who offered to purchase the Mersey & Irwell Co.
27 T. Lowndes, *The History of Inland Navigations, particularly that of the Duke of Bridgewater &c.* (3rd edn., 1779), 89, quoting a letter of 22 September 1767.
28 Stamford also supported the duke in his parliamentary campaigns for improving the waterway system.
29 Banks, 'Journal', *op. cit.* Punctuation, which he did not use, has been inserted in all these extracts.

30 *Ibid.*, 86.
31 *Ibid.*, 88.
32 *Ibid.*, 92.
33 *Ibid.*, 93.
34 *Ibid.*, 94.
35 *Ibid.*, 105.
36 *Ibid.*, 101.
37 *Ibid.*, 103.
38 *Ibid.*, 85.
39 J. Sharp, *Extracts from Mr Young's Six Months' Tour*, *op. cit.*, 23.
40 He was still in Birmingham in 1765.
41 H. W. Dickinson, *Matthew Boulton* (Cambridge, 1937), 50–1.
42 SEM, J. to T. Gilbert, 25 November 1767.
43 *Ibid.*

Chapter 10. The Trent and Mersey

1 F. Mathew, *The Opening of Rivers for Navigation* (1655), quoted in W. T. Jackson, *The Development of Transportation*, 187.
2 SEM, Rev. John Fenton to J. Loch, Stoke Lodge, 19 December 1843.
3 W. T. Jackman, *The Development of Transportation*, *op. cit.*, 186 ff.
4 For the Taylor and Eyes survey *cf.* J. Aikin, *A Description of the Country . . . Around Manchester*, 117; J. Phillips, *A General History of Inland Navigation*, 4th edn., 114–15; T. C. Barker, 'The Sankey Navigation', TLC, vol. 100 (1948), 143, giving further references.
5 SEM, Rev. J. Fenton to J. Loch, 19 December 1843. This confirms the view of V. I. Tomlinson in 'Salford activities', *op. cit.*, 63, that the duke was planning the Manchester link with his own canal by 1758. In fact he and Gower and Anson were also evolving an arterial system.
6 SH, Sec. 5, 5.
7 CRO Northants, EB 1460, 28 October 1766.
8 SEM, Petition for the Stockport Act. Also, *The History & Directory of Macclesfield* (Manchester, 1825), 49–52.
9 SEM, Edward Byrom and others to Lord Strange, 9 February 1765.
10 SEM, Smith & Allen, solicitors, Clements Inn, London, to Allen Vigor, Attorney, Manchester, 21 January 1766.
11 W. H. Chaloner, 'Charles Roe of Macclesfield', TLC, vol. 62 (1950–51), 146.
12 T. C. Barker, 'Lancashire coal, Cheshire salt and the rise of Liverpool', *Trans. of the Historic Society of Lancashire & Cheshire* (1951), vol. 103, 83 ff.
13 For further details *cf.* Chaloner, 'Charles Roe', *op. cit.*
14 Josiah Wedgwood to John Wedgwood, 6 July 1765, Wedgwood Museum, Barlaston.
15 See my *The Canal Duke*, 1st edn. (1961) for further details.
16 E. Meteyard, *Life of Josiah Wedgwood*, vol. I, 365.
17 Chaloner, 'Charles Roe', *op. cit.*, 150.
18 *Ibid.*

19 Chaloner, *op. cit.*, 154–5.
20 E. Meteyard, *Life of Josiah Wedgwood, op. cit.*, 420.
21 Meteyard, *op. cit.*, 420.
22 Meteyard, *op. cit.*, 420 ff.
23 J. R. Harris, 'Liverpool canal controversies', *op. cit.*, 1769–72, 1717c.
24 SEM, George Merchant to the Mayor of Liverpool, 17 December 1768.
25 SEM, J. Brindley to J. Gilbert, 21 December 1765. Brindley's spelling has been partly modernised.
26 L. A. Edwards, *Inland Waterways of England and Ireland* (1962), 340.
27 J. R. Ward, *The Finance of Canal Building in Eighteenth Century England* (1974), 29. I am most grateful to Professor W. H. Chaloner for this information.
28 CRO Northants, General accounts, EB 1460, 27.
29 *Ibid.*
30 CRO Herts, AH 1919. The duke's map 'for opening a communication between . . . the ports of Bristol, Liverpool and Hull'.
31 CRO Northants, *General State, op. cit.*
32 Some details of the canal debt are given in Appendix F.
33 SS, 467 (1862 edn.).
34 H. P. White, *Terrain, Technology and Transport History* (Salford, 1971), 3, and map of *Spillways*.
35 Malet, *History Today*; Brindley, *op. cit.*, 273.
36 A similar view was held by the Rev. F. H. Egerton.
37 J. Wedgwood to T. Bentley, 26 September 1762, Wedgwood Museum, Barlaston.
38 I am grateful to Charles Hadfield for pointing out that these were the Droitwich (1771), the Staffordshire and Worcestershire (28 May 1772) and the Birmingham Canal (14 September 1772) – this excludes the first part of the Bridgewater. Also, C. Boucher, *James Brindley, Engineer* (Norwich 1968), 114. But Brindley must have worked under J. Gilbert on the Staffordshire and Worcestershire.

Chapter 11. The battle of Norton Priory

1 CRO Herts, seen 1960, but not now calendared.
2 I am grateful to Mrs E. Brickell (*née* Malley), who has placed this interesting artefact on indefinite loan to the University of Salford. Also to Professor H. P. White for advice on early railway tickets.
3 CRO Northants, General State, *op. cit.*
4 *CNW*, 35 ff.
5 T. S. Willan, *River Navigation in England, 1600–1750* (1964), 58.
6 Not a misspelling but a *double-entendre* on his cousins' habit of varying the name Leigh. It is not known to which branch of that ancient family the duke is referring here.
7 SEM Cal., Duke to J. Gilbert, Ashridge, 16 June 1769.
8 CRO Herts; AH 1669–1670, show the process with the Canal Commissioners.

9 SEM, Draft petition by the Duke to Parliament against restrictions in Sir R. Brooke's estate, 6 March 1770.
10 *Ibid.*
11 Sir Henry Cavendish, *Debates of the House of Commons*.
12 CRO Herts, 12 July 1771 (not now calendared).
13 CRO Northants, General State, *op. cit.*
14 *CNW*, 34.
15 EM, 33.
16 T. C. Barker and J. R. Harris, *A Merseyside Town in the Industrial Revolution* (1954) 15–23.
17 A. E. Musson and E. Robinson, 'Science and industry in the late eighteenth century', *Economic History Review*, 2nd ser., vol. XIII, No. 2 (1960).
18 In SEM, the Gilbert correspondence.
19 R. M. Larking, *The Canal Pioneers*, p. 44. Mr Larking is a descendant of John Gilbert, and this work contains a valuable family tree.
20 J. Wedgwood to Thomas Bentley, 21 June 1773, Wedgwood Museum, Barlaston.
21 SEM, R. Brooke to the duke, from St James's, London; draft reply from the duke, Cleveland House, London, 29 January 1774.
22 SEM, Duke to R. Hill, 29 October 1775.
23 John Aikin, *England Delineated* (1788), 76 ff.
24 *Ibid.*

Chapter 12. Man with a mission

1 J. Wedgwood to T. Bentley, 23 May 1767, Wedgwood Museum.
2 Quoted in SEM general correspondence, from M. Dupin's *Force commerciale de la Grande [Bretagne]* (*c.* 1824). Dupin visited Worsley.
3 SEM, J. Benson to J. Loch, 25 December 1843. Benson thought that the date of this meeting was 1797.
4 A. Young, *A Six Months' Tour through the North of England*, vol. III, 1st edn., 1770.
5 EM, 119.
6 EM, 122.
7 SS, 406 (1862 edn.).
8 CRO Northants, General State.
9 J. Loch, quoted in SH – but Frank Mullineux has pointed out to me that Bradshaw also had land interests at Ashridge.
10 SEM, T. to J. Gilbert, 7 November 1776.
11 EM, 58–9.
12 SEM, T. to J. Gilbert, Navistock, 29 June 1778.
13 EM, 61.
14 EM, 63.
15 *Ibid.*
16 SEM, T. Gilbert to his father, J. Gilbert, Liverpool, 8 August 1776.
17 SEM, J. Gilbert to his son, Thomas, Worsley, 26 July 1778.
18 SEM, Mr Robins to J. Gilbert, Liverpool, 27 August 1789.

19 SEM, Duke to Mayor of Liverpool, 13 April 1790.
20 SEM, Mayor of Liverpool to the duke, 12 April 1790.
21 J. R. Harris, 'Liverpool canal controversies', 171, also states that the corporation opposed the widening of the Trent and Mersey canal.
22 J. Aikin, *England Delineated*, *op. cit.*, 79 ff.
23 Joseph Brotherton, first MP for Salford, 1832 to his death in 1857, made this statement in the Commons. *Cf.* C. P. Hampson, *Salford Through the Ages* (Manchester, 1930), 255 ff.
24 CRO Herts, 22 January 1782.
25 SEM, R. Lansdale to J. Loch, Worsley, 21 December 1843.
26 CRO Northants, General State, EB 1461, 1786.
27 EM, 101. There were then 207 colliers and 124 drawers.
28 CRO Northants, General State, EB 1461, 1786.

Chapter 13. Direct rule

1 MLH, *Lord Granville Leveson Gower, First Earl Granville, 1773–1846, Private Correspondence*, 1781–1821, ed. Castalia, Countess Granville, vol. 1, 1916, 70–2.
2 *Ibid.*
3 *Ibid.*
4 R. M. Larking, *The Canal Pioneers, op. cit.*, 39.
5 The evidence that the Mersey and Irwell offered to sell out to the duke is strong, if circumstantial – F. H. Egerton stated clearly in *The First & Second Part of a Letter*, p. 40, the reasons why the duke rejected this offer. V. I. Tomlinson, in 'Salford activities', states, p. 77, that the duke secured virtual control of the Salford Quay Company by deed of 1 May 1762. But later the Trustees purchased the Mersey and Irwell Navigation.
6 V. I. Tomlinson, 'Salford activities', *op. cit.*
7 EM, 12, mentions that 'a few years later', *c.* 1770, the duke owned 138 of the 200 shares of the Salford Quay Co.
8 MSC, Bridgewater Department, Archives. Mersey and Irwell Minute Book.
9 *CNW*, 267–8; V. I. Tomlinson, 'Salford activities', 9–10. It seems a little hard to condemn the duke's character because some Rochdale Canal minutes show that he bargained strictly with that waterway. *Cf.* James Beckett, *A History of the Rochdale Canal* (1956), NP, MLH, 6 ff. Also 34 Geo. III, cap. 78.
10 *CNW, loc. cit.*
11 *CNW*, 270 – not John Rennie, as is generally supposed.
12 *CNW*, 126 ff.
13 W. Wallace, *Alston Moor* (Newcastle upon Tyne, 1890), 124.
14 *Ibid.*
15 I am grateful to Frank Mullineux for this information.
16 C. Elsie Mullineux, *Mast and Pannage* (Swinton, 1964), 50.
17 C. E. Mullineux, *Pauper and Poorhouse* (Swinton, 1966), 7; SEP, R. Lansdale to J. Loch, 21 December 1843.
18 SEM, R. Statham to Mr Taylor, 12 February 1790.
19 I am grateful to the Rev. M. E. S. Meanley for his help. There is a tablet in the

chancel to Robert, a grave of his eldest son, John, and baptismal records for all three children.

20 SEM, R. Fenton to J. Loch, 19 December 1843.
21 *Ibid.*
22 Birmingham Public Library archives, box 4G.
23 I am grateful to Frank Mullineux for informing me that John Gilbert stayed on for a while at the New Hall. His mother left the Hall in 1796.
24 Not aged seventy-nine, as his memorial tablet states. I am grateful to W. Hoard Williams for this correction.
25 SEM, J. Fenton to J. Loch, 19 December 1843.
26 SEM, Duke to Mr Bury, 28 March 1797
27 SEM, Duke to Mr Bury, Cleveland Court, 7 March 1798.
28 SEM, Duke to Mr Varey, Cleveland Court, 27 November 1797.
29 Charles Hulbert, *Memoirs of Seventy Years of an Eventful Life . . .*, letters to his son, printed privately (Shrewsbury, 1852), 81.
30 C. Hulbert, *Memoirs, op. cit.*, 82.
31 Hulbert, *op. cit.*, 82, describes how he witnessed the scene from the far bank.
32 CRO Northants – the account books give details of coal carrying.
33 EM, 16.
34 J. H. Clapham, *An Economic History, op. cit.*, 78.
35 C. Hulbert, *Memoirs, op. cit.*, 59–60.
36 *Ibid.*
37 *Ibid.*
38 F. Mullineux, 'The Duke of Bridgewater's underground canals at Worsley', *op. cit.*, 155.
39 F. H. Egerton, *Transactions of the [Royal] Society of Arts*, vol. 18 (1800), 278.
40 F. H. Egerton, in *Transactions, op. cit.*, 278; but A. Rees, *Cyclopaedia*, states that work on the plane began in 1793 under J. Gilbert.
41 SEM, R. Lansdale to J. Loch, 21 December 1843.
42 *Ibid.*
43 F. H. Egerton, *op. cit.*, 269.
44 *Ibid.*, 275.
45 *Ibid.*, 276.
46 *Ibid.*, 269.
47 Manchester Ship Canal Company, Bridgewater Department archives; Mersey and Irwell Minute Book, 5 November 1794.
48 J. A. Rogerson, *AEI News*, vol. 19, No. 6 (June 1949), 8–9; Chetham's Library, Phelps collection, Worsley parish magazine article, 'The world's first steamboat (an incorrect claim)', by B. Chapman.
49 CRO Herts, AH 1914.
50 B. Chapman, *op. cit.*, 18; V. I. Tomlinson, 'Salford activities', TLC, *op. cit.*, p. 84, describes several local steamer experiments.
51 SEM, R. Lansdale to Mr Readman, Worsley, 11 Janauary 1844.
52 SEM, *ibid.*, 24 January 1844.
53 I am grateful for the view of Randall J. Le Boeuf Jn., of New York, that Fulton's contribution was probably very limited.

54 George Gower, Marquis of Stafford, 1803, first Duke of Sutherland in 1833.
55 SEM, J. Fenton to J. Loch, 19 December 1843.
56 F. H. Egerton, 'Underground inclined plane executed at Walkden', *op. cit.*, in *Transactions*, RSA, 265.

Chapter 14. Death of a duke

1 E. Richards, *The Leviathan of Wealth* (1973), 46. Expert opinion varies, but three hours for flats seems to be too short at Runcorn.
2 HLRO, JHC, 19 and 27 February 1795.
3 HLRO, JHL, 28 April 1795.
4 SEM. The Duckenfield purchase is dated 14 January 1799.
5 F. C. Mather, *After the Canal Duke* (1970), xvii.
6 CRO Northants, *General State*, EB 1461, 1803.
7 CRO Herts, AH 1834. I am indebted to Dr R. B. Schofield for information about the work on surveying.
8 SEM, Extracts from the Farington diary by Joseph Farington, R.A.
9 *National Gallery of Scotland, Shorter Catalogue* (n.d.), viii.
10 *Ibid.*
11 SEM, Sale catalogue of the Orleans collection.
12 Chetham's Library, Phelps collection, letter by W. Bentley Capper; SEM, Extracts from the Farington diary, *op. cit.*
13 CRO Herts, AH 1834.
14 The fine set of dining room (commode) chairs formerly in the Bridgewater Department's office at Castlefield, Manchester, were relics of these long dinners, at which much of the mining and canal business was transacted. Brindley's chair belongs to a private owner living in Salford.
15 SEM, Extracts from the Farington diary, *op. cit.*
16 SH, Sec. 7, 22.
17 CRO Herts, 16 December 1799.
18 I am grateful to the present owners, Mr and Mrs Arthur Lucas, for their hospitality, and for showing me the relevant archives and the Arkley Hole. The deed of sale was signed on 24 November 1801.
19 CRO Herts, D/EX 3/1–13. Carrington's diary runs from 1798 to 1810.
20 W. H. Chaloner, 'John Phillips: surveyor and writer on canals', *Transport History*, vol. 5, No. 2 (July 1972), 169.
21 CRO Herts, AH 1925–6.
22 W. H. Chaloner, 'John Phillips', *op. cit.*, 170.
23 CRO Herts, J. Cussans, *History of Hertfordshire* (1874–78).
24 *The Times*, 17 February 1802.
25 SEM, R. Lansdale to J. Loch, 21 December 1843.
26 *Ibid.*
27 British Museum, *Will of the third Duke of Bridgewater*, printed 1836, 66 pp.
28 SH, Sec. 1, 3.
29 F. C. Mather, *After the Canal Duke*, 335–6.
30 SH, Sec. 7, 26.

31 Just how busy and hard-riding a man the duke was can be seen from his coachman's accounts, CRO Herts, AH 1789–1819.
32 SH, Sec. 11, 2.
33 F. C. Mather, *op. cit.*, p. 1, estimates about 3,000 on the direct payroll of the trust alone by 1837.
34 F. H. Egerton, 'Francis Egerton, Third Duke of Bridgewater' (1809), quoted in V. I. Tomlinson, 'Salford activities', *op. cit.*, 86.
35 SH, Sec. 5, 7.
36 EM, *op. cit.*, 159. Mrs Brickell, *née* Malley, re-emphasised this point to the author in 1974, questioning whether there were any reliable sources for a contrary view.
37 W. H. Chaloner, 'James Brindley . . . and his remuneration as a canal engineer', TLC, *op. cit.*, 187.
38 *Annual Register*, 1803, *in* SH, Sec. 14, 1. Kippis gave a different time which is probably inaccurate.
39 *The Times*, 17 March 1803. I am grateful to the Duke of Sutherland for permission for Moira Leggat to photograph this mask, and other works in his collection.
40 *Annual Register*, 1803 *in* SH, Sec. 14, 2–3.
41 SH, Sec. 7, 20.
42 *Ibid.*
43 SH, Sec. 3, 2.
44 SH, Sec. 3, 3.
45 SH, Sec. 11, 1.
46 SH, Sec. 11, 3.
47 F. C. Mather, *After the Canal Duke*, chapter 1; E. Richards, *The Leviathan of Wealth* (1973), 43.
48 E. Richards, *The Leviathan of Wealth*, 143.
49 F. C. Mather, *op. cit.*, 9.
50 *Ibid.*
51 This has been admirably established by Eric Richards, *The Leviathan of Wealth, op. cit.*
52 E. Richards, *op. cit.*, 10.
53 *Ibid.*
54 SH, Sec. 12, 2.

Chapter 15. The aftermath

1 SH, Sec. 12, 1. I am most grateful to F. C. Mather for pointing out that the duke's will suggested William Rigby Bradshaw, not Captain James Bradshaw, R.N., as successor. Holme got the brothers mixed up.
2 C. E. Mullineux, *Pauper and Poorhouse* (Swinton, 1966), 19; *Victoria County History of Lancashire*, vol. IV, 381.
3 Mullineux, *Pauper and Poorhouse*.
4 SH, Sec. 12, 2.
5 SH, Sec. 12, 3.
6 F. C. Mather, *After the Canal Duke*, 320.

7 John McVicar to Lord F. Egerton, 25 November 1835.
8 T. Campbell to F. Egerton, 28 October 1843.
9 *Victoria County History of Lancashire*, vol. IV, 381.
10 *Morning Post*, in Mullineux, *op. cit.*, 19.
11 F. C. Mather, *op. cit.*, 329.

Bibliography

Manuscript sources

Documents relating to the duke's achievement suggest that the Egerton estates were so widely dispersed partly as a result of inheritance, and partly perhaps, from design. They reflect the English system of primogeniture, and were centred on Ashridge in Hertfordshire. The present location of archives is listed below in order of relevance to the third duke's life and work.

SH. Located at Mertoun, Roxburghshire, seat of the sixth Duke of Sutherland – Strachan Holme, librarian at Bridgewater House, London, *c.* 1920, set out to write *The Family Chronicles*, a gathering of notes and extracts leavened with his own assessments, on this branch of the Egerton family. *The Life of the Most Noble Francis Egerton* ... is a sub-section of this, and the most valuable single document for the duke's story, though somewhat limited by his not carrying out much research at Worsley, or elsewhere. It forms, as far as it goes, a reasonably detached and unbiased survey of some of the duke's activities, but the family did not then think it right to make Wood's private reports on the Grand Tour available. The twelve sections relating to the duke's life are:

1 Introduction; Life of the most Noble Francis Egerton ...
2 The Chancery suit against his mother and others.
3 His relations with General J. W. Egerton, F. H. Egerton (including F. H. Egerton's *Bridgewater Anecdotes*).
4 His relations with Thomas and John Gilbert and with Brindley.
5 His links with canal making generally, and with J. Wedgwood.
6 Extract from Collins's *Peerage*.
7 Francis Egerton, Lord Ellesmere's 'Essays on History, Biography &c.' contributed to the *Quarterly Review*, 1858. (Holme regarded this as the best of the altogether inadequate summaries of the duke's life.)
8 Extract from *Town and Country Magazine*, 1775.
9 His racing.
10 Roman relics at Castlefield.

11 The duke's will and the Bridgewater Trust.
12 The Superintendents.
13 Transfer of management from Superintendent to Chief Agent.
14 Obituary notices.

SEM. Sutherland family Estate papers, Mertoun. Most of the archives at Bridgewater House were calendared – listed and typed with brief bound summaries – but very few of the third duke's papers were included, and many seem to have reached Mertoun after the calendaring. Of prime importance are letters written to J. Loch at Lord Francis Egerton's request, to gather material for his 'Essays on History, Biography &c.' (Sec. 7 in SH) from Robert Lansdale and the Rev. J. Fenton. The Gilbert letters – some only in draft form – offer valuable insight into the duke, the Gilberts and other Bridgewater agents at work. The varying collection of other papers include most of the receipts, etc, for land taken over for the canal as it progressed.

CRO Northants. County Record Office, Northamptonshire. Contains the vital 'General State of His Grace the Duke of Bridgewater's Navigation, Colliery, Lime & Farm Concerns in Lancs. & Cheshire from Midsummer 1759' kept by John Gilbert; this shows the pyramid as the debt built up to 1787, and the decline as it began to be paid off. Some of the account books which fed this key work with information include *The Day Book of the Duke of Bridgewater* (title page now missing), and EB 1460, entitled *D No. 1, General Accounts* from 1759 to 1790, the first thirty-six pages of which, to 12 September 1767, are possibly in the duke's own hand and are certainly audited and signed by him. The first seven pages are left blank for earlier work which was not entered. Of lesser interest are William Brough's general expenses with the duke from 1766–68, and Mathias Shelvoke's similar accounts from 1769–70.

EM. Edith Malley (Mrs E. Brickell), 'The financial administration of the Bridgewater estate, 1780–1800', an unpublished MA thesis for Manchester University, 1929. Though this refers only to the north-western area of the duke's estates, it is a most important survey of his transport, mining and agricultural activities, drawn from archives at the Bridgewater offices, Walkden, which can no longer be traced. It seems that they were burned on nationalisation.

CRO Herts. County Record Office, Hertfordshire, containing the Ashridge House – AH – collection, valuable for showing the duke, the Gilberts and the Tomkinsons in action, the bonds, or loans used for financing the canal and mines, the duke's canal campaign maps and receipts for his paintings, charity, etc.

BES. Bridgewater Estate Co. archives, formerly at Worsley. This collection included Massey's account books. Present whereabouts uncertain. It also includes letters to and from Sir Richard Brooke relating to the Norton Priory problem.

HLRO. House of Lords Record Office. Journals of the Lords and Commons are vital for assessing the part played by the duke and his agents in gaining legislative support for the arterial canal idea, and contain valuable incidental information.

Other archive sources include:

Bridgewater Department, Manchester Ship Canal Company, Mersey & Irwell Navigation Co., Minute Books. Chethams Library, Manchester, the Phelps collection – a local historian's garnering of information relating to the Egertons and Worsley. CRO Bedford – material on the Ashridge estate. Manchester Central Library, Local History Department, valuable publications and cuttings. Birmingham Public Library, Boulton & Watt collection, box 4G, includes letters from John Gilbert Jr.

Directories and newspapers of value for this period include: Mrs Raffald's *Directory*, Lewis's *Directory of Manchester & Salford*, 1788, and *The Manchester Mercury*. The Salford Central Library, Peel Park, and Chetham's Library, Manchester both contain important collections.

Among the more remarkable discoveries of archives in recent years was a cache of documents at the author's house in Cheshire. While rewiring under the roof there Mr M. Costello uncovered bundles of papers and parchments dating back to 1698, which included five freight bills for John Antwiss in account with 'His Grace the Duke of Bridgewater', for corn and meal carried from Bartington Farm wharf on the adjacent Trent & Mersey canal, from 1792 to 1795. At present these are the only known invoices surviving from the Bridgewater days; it is significant that they are accounted direct to the duke in person, rather than to agents or a company.

A very recent acquisition of the Buile Hill Mining Museum at Salford is the *Account Book For Walkden Moor Colliery, January 1781–December 1816*. This is Account Book No. 25, which covers payments for extending and renewing the Old Sough, or higher navigable level, and, in considerable detail, the mining costs and coke oven expenses.

Further reading

In the following list, London is normally the place of publication, unless otherwise stated. (The bibliography of Lancashire canals by M. W. Moss, which is listed below, is available in MLH, and at the Library Association, London.)

Aikin, J. *England Delineated* (1788)

— *A Description of the Country from thirty to forty miles around Manchester* (1795, reprinted 1968)

Anon. *Remains Historical & Literary Connected with the Palatine Counties of Lancaster & Cheshire*. Vol. 42 (1899)

Anon. (Manchester Ship Canal Co. publicity). *The Bridgewater Canal – Bi-centenary Handbook* (1961)

Anon. *The Constables' Accounts of Manchester, 1612–1776* (Manchester, 1891)

Ashmore, Owen. *The Industrial Archaeology of Lancashire* (Newton Abbot, 1969)

Ashton, T. S. *The Industrial Revolution, 1760–1830* (1948; reprinted 1970)

Bagley, J. J. *Historical Interpretations*. Vol. 2 (Newton Abbot, 1972)

Banks, A. G. and Schofield, R. B. *Brindley at Wet Earth Colliery* (Newton Abbot, 1968)

Barker, T. C. and Harris, J. R. *A Merseyside Town in the Industrial Revolution – St Helens, 1750–1900* (1959)

Bell, S. Peter, ed. *Victorian Lancashire* (Newton Abbot, 1974)

Body, A. H. *It Happened around Manchester – Canals and Waterways* (1969)

Boucher, C. T. G. *James Brindley, Engineer* (Norwich, 1968)

Bridgewater. *Will of the Duke of Bridgewater* (1836)

Burton, A. *The Canal Builders* (1972)

Chalmers's *Biographical Dictionary*, Egerton (1814)

Chaloner, W. H. *People and Industries* (1963)

Chapman, W. *Observations on the Various Systems of Canal Navigation* (1797)

Clapham, J. H. *An Economic History of Modern Britain*. Vol. 1 (Cambridge, 1950, reprinted 1964)

Clifford, F. *A History of Private Bill Legislation*, 2 vols. (1885, reprinted 1968)

Clutterbuck, R. *History and Antiquities of the County of Hertford*, n.d. (in CRO Herts).

Curwen, S. *Journal and Letters* (1842)

Cussans, J. *History of Hertfordshire* (1874–78)

Dickinson, H. W. *Mathew Boulton* (Cambridge, 1937)

Edwards, L. A. *Inland Waterways of England and Ireland* (1962, reprinted)

Egerton, Rev. F. H. *The First and Second Part of a Letter upon Inland Navigation* (Paris, n.d.)

Éspinasse, Francis. *Lancashire Worthies* (1874)

Falk, B. *The Bridgewater Millions* (1942)

Fournel, Henri and Dyèvre, Isidore. *Mémoire sur les canaux souterrains de Worsley* (Paris, 1842)

Gibbs, Vicary. *The Complete Peerage* (1916)

Granville, Castalia, ed. *Lord Granville Leveson-Gower, 1774–1846, Private Correspondence, 1781–1821*, vol. 1 (1916)

Hadfield, Charles. *The Canal Age* (Newton Abbot, 1968; reprinted 1971)

— *British Canals* (Newton Abbot, 1950; reprinted 1969)

— *The Canals of the West Midlands* (Newton Abbot, 1966; reprinted 1969)

Hadfield, Charles and Biddle, Gordon. *Canals of North West England*, 2 vols. (Newton Abbot, 1970)

Hart Davis, Henry V. *The History of Wardley Hall* (1908)

Holme, Strachan. *FAMILY CHRONICLES. The Life of the Most Noble Francis Egerton.* NP, ND (*c.* 1920)

Hughes, S. *Quarterly Papers on Engineering*, vol. 1 (1844)

Hulbert, Charles. *Memoirs of Seventy Years &c.* (Shrewsbury, 1852)

Jackman, W. T. *The Development of Transportation in Modern England* (1916; reprinted 1962 and 1966)

Larking, R. M. *The Canal Pioneers* (published privately 1967)

Leech, Sir Bosdin. *The History of the Manchester Ship Canal*. 2 vols. (1907)

Lowndes, T. *The History of Inland Navigations, Particularly . . . of the Duke of Bridgewater in Lancashire & Cheshire* (1766; reprinted 1769)

Malley, Edith. 'The financial administration of the Bridgewater estate, 1780–1800' (unpublished M.A. thesis, Manchester University, 1929)

Macclesfield. *The History and Directory of Macclesfield* (Manchester, 1825)

Mather, F. C. *After the Canal Duke* (1970)

Mathew, Francis. *The Opening of Rivers for Navigation* (1655)
McCutcheon, W. A. *The Canals of the North of Ireland* (Newton Abbot, 1965)
Meteyard, E. *Life of Josiah Wedgwood*. 2 vols. (1865–6)
Moss, M. W. 'Canals of Lancashire, 1700–1900: a bibliography and guide to the literature' (MS, copies in MLH and Lib. Assn, London 1973)
Mullineux, C. E. *Mast and Pannage* (Swinton, 1964)
— *Pauper and Poorhouse* (Swinton, 1966)
Mullineux, F. *The Duke of Bridgewater's Canal* (Eccles, 1959; reprinted 1970)
— *The Duke of Bridgewater's Underground Canals at Worsley* (Eccles, reprinted from TLC, vol. 71, 1961)
Musson, A. E. and Robinson, E. *Science and Technology in the Industrial Revolution* (1969)
Nef, J. U. *The Rise of the British Coal Industry*. 2 vols. (1966)
Ogden, James. *A Description of Manchester* (Manchester, 1783)
Owen, David. *Water Highways* (1967)
Phillips, J. *A General History of Inland Navigation* (1785; reprinted 1803)
Plumb, J. H. *England in the Eighteenth Century* (1950; reprinted 1961)
Rees, Abraham. *The Cyclopaedia or Universal Dictionary*, vol. 16 (1819)
Richards, E. *The Leviathan of Wealth* (1973)
Rolt, L. T. C. *The Inland Waterways of England* (1950; 5th edn., 1970)
— *From Sea to Sea* (1973)
Sharp, James. *Extracts from Mr Young's Six Months Tour Through the North of England* (1774)
Sharpe France, R. *Guide to the Lancashire Record Office* (Preston, 1962)
Smiles, Samuel. *Lives of The Engineers*, vol. 1 (1861; reprinted 1904)
Smith, John H., ed. *The Great Human Exploit* (1973)
Tate, W. E. *The Parish Chest* (Cambridge, 1946)
Thompson, Colin and Brigstocke, Hugh, eds. *National Gallery of Scotland – Shorter Catalogue* (Edinburgh, 1970)
Victoria History of the County of Lancaster (1906)
Wadsworth, J. P. and de Lacy Mann, J. *The Cotton Trade and Industrial Lancashire* (Manchester, 1931; reprinted 1965)
Wallace, W. *Alston Moor* (Newcastle upon Tyne, 1890)
Walpole, Horace. *The Yale Edition of Horace Walpole's Correspondence* (1937 onwards)
Ward, J. R. *The Finance of Canal Building in Eighteenth-century England* (1974)
Weale, Johan. *Quarterly Papers on Engineering*, vol. 1 (n.d., MLH)
Wheat, G. *On the Duke's Cut* (Glossop, 1977)
Willan, T. S. *River Navigation in England, 1660–1750* (1933; reprinted 1964)
Williams, A. J. 'The impact of the Bridgewater Canal on land use' (unpublished M.A. thesis, Manchester University, 1957)
Young, Arthur. *A Six months' Tour Through The North of England*, vol. 3 (1770).

Articles of particular relevance

Chaloner, W. H. 'Francis Egerton, third Duke of Bridgewater (1736–1803): a bibliographical note'. *Explorations in Entrepreneurial History*, vol. 5, No. 3 (March 1953). Research Notes; 181

Clegg, H. 'The third Duke of Bridgewater's canal works in Manchester'. *TLC*, vol. 65 (1955), 91–103

Egerton, F. H. 'The Inclined Plane'. *Transactions of the Society of Arts*, vol. 18 (1800), 265–79

Hassall, E. R. and Trickett, J. P. 'The Duke of Bridgewater's underground canals'. *Mining Engineer*, 37 (October 1963), 45–57

Malet, H. 'Brindley and canals'. *History Today*, vol. 23, No. 4 (April 1973), 266

— 'The Duke of Bridgewater and the eighteenth-century energy crisis'. *Journal of the Royal Society of Arts*, vol. 123, No. 5226 (May 1975), 374

Mullineux, F. 'The Duke of Bridgewater's underground canals at Worsley'. *TLC*, vol. 81 (1961), 158

Mortimer, John. 'John Varey's Cash Book.' Papers of the Manchester Literary Club, vol. xxi (1895), 53–69. The reader should guard against the inaccurate statements on p. 56, relating to the underpayment of Brindley by the duke. Late payment of salaries was also more frequent in the eighteenth century than now.

Tomlinson, V. I. 'Salford activities connected with the Bridgewater Canal'. *TLC*, vol. 66 (1956), 51–86

Index